# CHRONICLES OF THE GLEN

## CHILDHOOD ANECDOTES AT POPLAR GLEN FARMS

Keith Erwin Brower

 FriesenPress

Suite 300 - 990 Fort St
Victoria, BC, V8V 3K2
Canada

www.friesenpress.com

**Copyright © 2017 by Keith Erwin Brower**
First Edition — 2016

All rights reserved.

No part of this publication may be reproduced in any form, or by any means, electronic or mechanical, including photocopying, recording, or any information browsing, storage, or retrieval system, without permission in writing from FriesenPress.

ISBN
978-1-4602-9515-1 (Hardcover)
978-1-4602-9516-8 (Paperback)
978-1-4602-9517-5 (eBook)

1. BIOGRAPHY & AUTOBIOGRAPHY, PERSONAL MEMOIRS

Distributed to the trade by The Ingram Book Company

Maris &

Happy Birthday!

All the Best

Keith
2019

# For my children

Jennifer Lynn, Rachelle Anne,
Jodi-Ellen, Mitchell
and their families

Thanks so much!

# CHAPTER 1

## Meeting the Bus

It was a cold day—not the kind of cold provided by a prairie blizzard, but more like the kind of cold experienced on a calm winter day when the sun sinks below the horizon in the late afternoon. I was a small child, out helping or tagging along with my father as he worked away at the daily chores that a mixed farm required. We spent a lot of time outdoors as kids; there were all kinds of things to explore, animals to play with, and hideouts to curl up in (like the oat bundle stacks). There were no screens in those days so we had to make our own fun, mostly outdoors.

My older brother was at school and was expected home soon. I looked forward to the time he arrived as it meant that there would be someone to join in the frolic of the evening. Interestingly, he was assigned to ride the Fabyan school bus; he was picked up and dropped off not in front of our farmyard but a mile west on the corner of the crossroads. Can you imagine a grade one or two child in today's world being asked to walk a mile to and from the school bus in the middle of winter on a lonely country road?

On the day in question, I asked my father if I could walk along the road to meet my brother. My father looked at his watch and agreed that the bus should be dropping him off soon. He carefully removed his watch, which was tied to his belt by a piece of leather thong split on the end, forming a slot so the watch could be slipped through, and long enough so that when he removed the watch from its special pocket he could hold it out in his hand and read the time easily. I was not familiar with the idiosyncrasies of telling time from such a timepiece so Dad carefully pointed out where the hands were, where they should be when the bus arrived, and their position when I was to return, even if the bus didn't come.

My father removed his watch and with it carefully placed in my pocket, my mitts adjusted, my snug winter coat fastened, and my warm moccasins on tight, I started off with all the confidence that a four-year-old can muster. And I was warm. My mother had ripped out some old hand-me-down garment from an adult connection she had and had cut and sewn a coat just my size that included a coyote-fur fringe hood. The coyote had been trapped and its hide tanned by my father. I wore breeks and felt-lined moccasins that had arrived earlier in the winter from the Army and Navy in Winnipeg. My mother managed to scrape enough cash together each fall to send an order in and it was always a highlight when it arrived in the mail.

I don't remember much about the walk to the corner but I did make it. I stood for a long time waiting for the bus, which never arrived. To this day I don't know whether I missed it or whether perhaps my older brother wasn't on the bus that day and therefore the normal route past the corner wasn't taken. At any rate, the day grew darker as time passed and I decided to remove my mitt and pull out the watch to check the hands. According to my father's instructions, it was time to start back and so I reluctantly prepared for the walk home. The watch was carefully placed back in my pocket and I began to try to put my mitt on. This is where the problems began. Try as I might, I couldn't pull the mitt onto my free hand using the bulky mittened hand.

As it grew darker I began the journey back to the farmyard. By now my uncovered hand was feeling cold and, being so young, I did not have experience in how to keep a hand warm when you don't have a mitt to cover it. I could have placed it between my arm and body, for example, but all I can recall is walking and holding onto the removed mitt and feeling the cold creeping into my entire hand. As the tingling turned to numbness, I began to cry. There I was, a small boy with a freezing hand and the impending gloom of darkness closing in on me. Who knew—maybe the coyotes came out after dark and ate little children. The very thought only added to my misery.

Keith Erwin Brower

I don't remember how long it took me to walk the mile home but I do remember being rushed to the sink by my mother, who got some snow and cool water to place my hand in and then gradually added warmer and warmer water from the hot water reservoir at the end of the kitchen stove. Even now I can feel the tingling as the feeling began to return to my hand.

I can imagine that my father would have received a lecture about small boys, cold days, and long treks on lonely country roads. That conversation I didn't hear nor was I a part of it. I do know that I never met the bus again.

# CHAPTER 2
## The Chicken Yard

There it was, in the middle of a tangle of box elders (Manitoba maples), raspberries, and poplars, amongst other various unkept vegetation. It was held up with barn hinges folded around and nailed to a charred post, still firm and unbending after all these years. For a gate, it wasn't wide: a few shiplap boards held together at the top and bottom with a cross member of similar material nailed to the uprights to fix it all in place. The top was cut so there was a peak at the centre and the latch hung loosely, no longer needed to hold the gate from swinging open.

In its prime the gate was at the northeast corner of the chicken and turkey yard. This was a rather large fenced-off area, where the poultry of the day could wander at will during the long days of summer, developing into adults to produce eggs or to be slaughtered for meat in the fall. The birds ranged all over this large enclosure and had "free range" of the whole area, long before "free range" was the latest buzzword. The blades of grass and clumps of weeds didn't stand a chance, and it was not an area where any worm or self-preserving bug dared wander either.

As I recall, every second post was charred, especially the corner posts. The others were bluestone-treated poplar saplings that were long enough to hold up the five-foot hexagonal poultry mesh. The poplar posts were developed by cutting straight young poplar trees in the spring. These were stood up in a half-barrel or tank filled with water that contained dissolved bluestone (copper sulphate). As the capillaries in the wood drew water up, the dissolved bluestone was drawn up with it and filled the wood with the bacteria-arresting effects of the copper sulphate. In this way a well-treated post lasted for many years.

As to the charred posts: they had been in a fire. Not at the farm but in the local Atlas lumberyard; they were the remainder of six-foot-by-eight-inch treated posts that were part of the inventory of the establishment when it went up in smoke. These were now six-inch posts, having lost their outer circumference to the flames. The centres were still solid however, and the remaining piles of these pickets were sold to locals at a large discount. Quite a number of these posts showed up at the farm, as did sheets of plywood that once measured four feet by eight feet but after the fire were three by six feet. Both commodities found many uses around the farmyard for a number of years.

The chicken pen enclosed two different small outbuildings. The small house for layers was a single-story shed-roofed affair with a single walk-in door. A nesting box area at waist level was across the east end with gunny feed sacks used as modesty covers. The hens ducked under these covers to enter the nests and lay their eggs. The roosting area was at the west end. It consisted of unpeeled poplar saplings two to three inches across, held in place with number nine wire. Under the roost a piece of the burned-out plywood served as a dropping catch. The building was capable of housing up to thirty-five hens, a goodly number that kept the family in eggs. The construction was wood-frame lined with shiplap. Between the exterior and interior walls, chopped straw was used as insulating material. There were a couple of threadbare power wires coming from the yard light pole that provided enough power for a light, which was especially appreciated in the long dark days of winter. A cord could be extended from it to the brooder next door to run the heat lamp.

A self-feeder was constructed and nailed to the wall. It was approximately ten inches wide at the top and tapered down to a couple of inches wide at the bottom, where it emptied into a feed box with a lip. Its depth was the same ten inches and the whole device was perhaps twenty-four to thirty inches wide. A steep angled cover with attached leather hinges and a handle covered the feeder except during filling. The angle was steep enough to prevent hens from roosting on it and leaving a mess. The construction of the building also lent itself to having other inhabitants. With the shiplap interior, which naturally had knotholes, and with the abundant feed, as well as the nice straw insulation, it was natural to have small inhabitants that very much enjoyed the setup. Quite often they were spotted scurrying here and there when you first stepped into the shed's interior. Heaven help the little mouse that was spotted by a hen. The hens could be very quick as they chased these little rodents but usually the mice were able to escape into a hole in the interior wood.

The second building was an old granary that had been moved into place to serve as a poultry shelter, possibly from my grandfather's yard a mile and a half to the east. As his farming career wound down he no longer had a need for all his granaries. At any rate, this building started in the spring as the brooder house and ended up as housing for the turkeys and roosters in inclement weather in summer and, if they were still with us, the early winter.

The spring always brought a shipment of chicken and turkey poults by Rail Express or mail. What a great time it was to have a box of exuberant chicks peeping away and peeking their small beaks through the many breathing holes that perforated the shipping container. They were always mixed up, usually in four different compartments with both turkeys and chickens in each one. A small amount of thin grass-like shavings covered the bottom, an absorbent bedding to keep the youngsters dry and give them good footing. What fun we had placing these small birds in our brooder in the granary. We were taught to catch and lift them gently out of their shipping container, squeezing just enough to keep them from escaping but loosely enough so as not to injure them. They were placed in a brooder my father had built that required a heat lamp under one end, had rounded corners in the living area to avoid overcrowding, and was covered with a lid over the heat lamp end and one-inch mesh on the other. This whole affair was lifted to waist height by four two-by-four posts in the corners. The walls were about ten inches high and the floor consisted of six-inch shiplap. The chicks stayed in this brooder until they were a few weeks old and then the brooder was removed and the chicks were moved down to the floor of the granary along with the heat lamp, which was hung in a canopy near the floor to continue to provide heat for the growing poults.

As spring turned into summer, these young poults moved from the little granary to being able to venture outdoors on sunny days. At last, as the season wore on, they ventured into the large enclosure as full-fledged poultry. The little birds grew quickly, as they ranged freely in the big yard and had their fill of farm-raised grains. It was apparent that the two little buildings had served their purpose well by providing cover and shelter during these development days. Many of the birds would be tasty dinners celebrating special occasions over the next few months and, of course, our family was well-supplied with delicious eggs for the many varied uses my mother found for them.

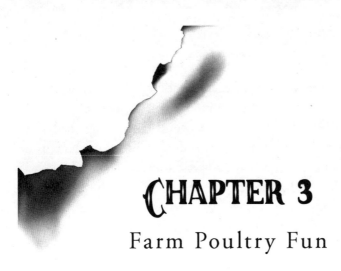

## CHAPTER 3

### Farm Poultry Fun

The layers were interesting to watch. In fact, if observed, so were all the farm animals that were present while I grew up. With slow movements and a quiet demeanour it was possible to observe them very closely; this was something that I did often as a child.

We mostly had dual-purpose hens, ones that were good layers and also good meat birds. Usually they were less flighty as well. One could go in the henhouse and stand quietly by the nest boxes and observe the actions of these feathered friends. They seemed very intelligent as they slowly walked around, leaning their heads this way and that. A slow walk with head erect, placing one foot in front of the other, first with claws folded and then extended, and a quiet, slow, but deliberate "bauck, bauck, bauck" meant all was well in the chicken yard or perhaps "I've just laid a fine egg." My grandmother said it was their way of singing. This action and these sounds were quite often present as one of the layers left the nest.

If you were very quiet and gentle you could gather the eggs under one of these girls by slowly placing your hand between her and the hay of the nest box. There you found the eggs under her, laid by her compatriots prior to her taking over the nest. On the very odd occasion, if you arrived just at the right time, she might lay the egg right into your hand. It was important to keep the hens happy, as there was a hungry family waiting for eggs' arrival and the various culinary delights that used them.

It didn't take much to upset the flock either. Any quick movement, whether a mouse, a door opening too quickly, a pail falling over, or a gust of wind blowing something by and the whole flock would be in panic mode. It started with one of the girls voicing "bluc, bluc, bluc, bla-douck, bluc, bluc, bluc, bla-douck," each phrasing ending in a high-pitched crescendo. Shortly the whole henhouse would join in the uproar and keep up the racket for a few minutes or until the flock perceived there was no longer danger. If there was something "twisting in the wind" that kept their attention, they could keep this up for quite a while.

Every spring a third interesting type of hen would arise. This was the "clucker." She was either sought-after or despised, depending on the circumstances. If, for instance, you happened to have some fertile eggs that needed brooding, then the clucker was just what you wanted. It didn't matter whether it was chicken eggs, turkey eggs, ducks, or geese, the clucker would hang in there to hatch any of them. It did become a little confusing when she hatched out ducks, for example, and she couldn't keep the ducklings out of the water. Talk about stress! On the other hand, if you were not in need of egg incubation, these old girls were a pain. They were forever in the nest, and when you reached in to get eggs usually gave you a throaty squaller and pecked your hand. That hurt a lot when you were little, and it had the desired effect: you left her alone. If she could be removed from the nest, she strutted around with her feathers puffed out and voiced her displeasure with a "cluck, cluck, cluck" wherever she went. Of course, when she took up brooding, she was no longer laying, and the longer on the nest with limited food intake, the more her flesh melted off and she wouldn't be good for chicken dinner either.

The turkeys were always very interesting as well. We raised both Nicholas Whites and Broad Breasted Bronze, depending on the year and supplier. The Broad Breasted Bronze were coloured like wild turkeys and the Nicolas were white (as their name suggests). I found them very cute and inquisitive. They would stretch their necks and try to peck things you held for them.

It seems that turkey poults were not as bright when it came to seeking out food and water in the early going. They were very attracted to shiny things and this was one of the reasons my parents used marbles in their waterer. When the turkeys pecked at the shiny marbles their beaks got wet and they realized this was their water source. The second reason for the marbles was to keep the turkeys out of the waterer's lip and thereby prevent them from getting wet and chilled. Usually when this happened the outcome was not good. As they grew, the yellow and brown of the pre-feather fuzz began to be replaced by coloured feathers (or the white feathers of the Nicholas). The hens began to put on size and would walk around with the "bauwk, bauwk, bauwk," typical of turkeys in a good comfortable situation. They would keep this up as they removed any spears of grass, weeds, or any other greenery that graced the yard. The toms, on the other hand, started puffing out their feathers and strutting around very early on. They would challenge us as youngsters, and we would respond. It was "be bop bop, be bop bop bop, bee bop bop," followed by a volley, "bbbthlalalalalal" and a snuffle through their nostrils. They seemed to be able to keep this up for hours. I often wondered when they ate. It was clear they did, as they weighed up nicely come the late fall; they were always very tasty for Thanksgiving and Christmas, as I recall.

Dad quite often had both ducks and geese as well. I remember being very young when I had my first introduction to the gander. All I remember is him coming at me with his wings spread wide, flapping and hissing like a whiteout. If they latched on to you they could pinch your flesh and leave marks. Not a nice sensation, and one that certainly made a small boy give the geese a wide birth whenever they crossed the yard. The goose eggs were very large in my hands. It took both to hold one. It was fun to accompany my father as he showed us the nest and how the goose had plucked down from her breast to line it with softness. Usually there were six to eight big white eggs and if all went well, in twenty-eight days a whole new group of goslings appeared. The geese were almost as good as guard dogs, as they could cause quite a ruckus when anything new entered the yard. Having them around was wonderful; the major drawback was having to watch where you stepped or lay down around the farmyard because of their bathroom habits.

The ducks we kept were usually Rouens, which had similar markings to the wild Mallard ducks. They usually were housed in the large chicken yard and had to be happy with a small amount of water as compared to the nearby pond. Unfortunately, although it would have been great for them to have had the freedom to go to and from the pond, there was always the sly opportunist to deal with. Many an anticipated duck dinner left in the mouth of one of the local coyotes, who were always in the shadows awaiting the chance to enjoy a well-fed tasty delight. Every once in a while the poultry yard had to be checked for holes in the fence; escaped fowl became a very real temptation to these canines.

My recollection is that not many exotic poultry species were present on the farm in those early years. I believe there were few guinea fowl once, but their ability to fly saw them quite often outside the poultry enclosure. Outside meant that they became fodder for the always-lurking coyotes and thus their presence in those early days was short-lived.

With this variety of poultry we always had wonderful meals, both light and dark meat. The aroma of one of these home-grown birds in the old oven roaster, mixed with the scent of seasoned poplar logs burning in the kitchen range, freshly baked buns, and one of my mother's freshly baked pies made us think we were one of the best-off families in the world. And we were. We ate very well as youngsters and certainly the poultry on the farm were a large part of that.

# CHAPTER 4
## Lexi and Flicka

Heavy horses are huge, especially when you are a small boy. The team on the farm when I was growing up consisted of a pair of mares, sisters, that were of the Belgian breed. Known as Belgians or Brabants, they were a light chestnut colour with flaxen manes and were good representatives of their breed. Belgians have tremendous power and are one of the largest heavy horse breeds. Mares often weight up in the 1600- to 1700-pound range and stand sixteen plus hands high. As I recall, Lexi and Flicka fit the bill as described for colour, height, and weight. I remember the echo of their large hooves on the cement floor in the centre of the barn as my father led them outdoors to be hitched up for the work of the day. Their feet reminded me of dinner plates as they pranced by. They were used sparingly, however, because by the time I had arrived on the farm scene, this kind of horsepower was being replaced by the mechanical type.

The horses were so interesting to watch. They would stand together and nuzzle and seemed to be chatting to one another as they stood awaiting instruction. It was interesting how they could shift weight from one hind leg to the other by relaxing one side; when tired of that stance, they would straighten that leg and relax the other. Another fascinating thing that amazed me was the way that they could ripple and wiggle their hide or skin to send flies on their way or deal with some other irritant. Naturally this was a full-time job as the flies kept coming back.

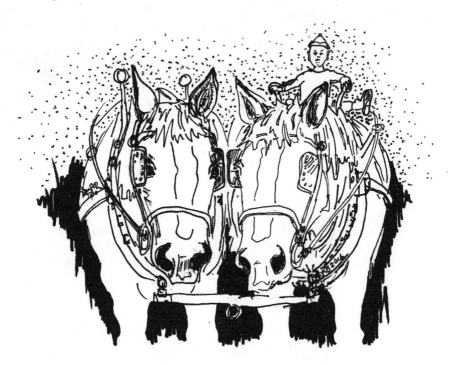

As I recall, they were gentle giants. They were so engaging as they ate their oats and sorted through the whole oat bundles provided them for roughage. Their very flexible upper lips were so agile that they could separate the provided roughage to devour the best morsels first. The smell of fresh-cut green and dried oats immediately reminds me of the team chewing and enjoying their evening meal. I can still hear the crunch as they chewed away.

I'm sure, as in any situation, one of the girls was in charge, probably Lexi, as she was a year older than Flicka. They mostly worked as a team; however, at times Flicka was driven in a single harness, to pull the stone boat, for example. This was a rectangular implement with runners down each side made of wood and covered with strap iron. The name says it all; at times it was used to throw stones onto when new cropland was being developed. The stone boat was then driven to a pothole where the stones were dumped and the whole process started again. It was also used in the winter as a manure spreader, loaded at the barn, driven to the field to be unloaded at a slow walk, and brought back to be reloaded. This continued until the job was done. The milk or cream cans from the dairy were sometimes pulled to the front gate by the stone boat when the yard was too full of snow to get them there any other way.

The team was also used on the hay rack fixed with sleight runners for the winter. My father loved kids and enjoyed having us along. My brother and I, all bundled up, would go for the mile-and-a-half ride when getting straw with a toboggan tied to the back of the rack. Sometimes when the snow was deep the ruts left by the horse prints and sleigh runners would flip us over and Dad would bring the outfit to a stop while we readjusted the toboggan and climbed back on. Often we would have a small fire built away from the straw stack to keep us warm while he loaded the rack before our return trip. Needless to say, he wasn't in need of being warmed after forking on the whole load of straw.

Dad could call the team from anywhere in the pasture by making three shrill whistles. The girls would stop what they were doing and raise their heads in anticipation. Three more and they were on their way to the barn. I'm assuming Dad would have some reward for them, like oats, when they arrived.

It must have been one of the last summers on the farm for the girls the year they were involved in making silage. Dad was one of the first people in the district to adapt his winter forage needs by putting up silage rather than dry hay. For this he needed to retrofit the wagon that had been used for grain and had the four-and-a-half-foot oak-spoked wheels wrapped with iron tires and large oak hubs. These had been changed to rubber; the spokes were cut down to fit inside the rim of a rubber air-filled tire, which was naturally much smoother-rolling than the steel rims and the oak wheels had been. The sides, which were approximately twenty-four inches high, had been extended with wooden boards so that a larger load could be blown in. This was done with the Brady forage harvester that followed along behind the 15-30 McCormick Deering, the farm's first tractor, acquired in the early 1950s. The harvester used flail-type chains rotating on a drum that cut and blew the crop into the trailing wagon. Its big disadvantage was that it was not side-mounted and therefore the tractor tramped down a lot of crop. Dad eliminated a lot of this by making a metal V that was mounted on the tractor fender, using a rod to lower the V to just below crop height. It spread the grain apart just before the wheel of the tractor caught it and thus a lot less crop was trampled. In the bottom of the wagon were two large ropes, one on each side. These were fastened firmly at the back and then laid on the floor toward the front of the wagon. Wooden slats lay between the ropes and were fastened to the ropes at the end of each slat. These slats continued all the way to the front where the leftover length of both ropes were hung and joined together to give one pull point.

This is where the girls fit in. The full wagon, forage harvester and all, were driven to the pit, where the loose ends of rope were thrown over the load and hitched to Lexi and Flicka. The command was given and they pulled their hardest and slowly the load began to roll up and out of the wagon. Once the rope and slat sling had rolled the whole load into the pit, it was reset into the bottom of wagon and the 15-30 was off again to repeat the process. Lexi and Flicka rested until the next load showed up.

I particularly remember one day during silage time. The day was starting and I was obviously out "helping" as the team was prepared for another day of pulling silage in the pit. It was a beautiful day, as I recall, with lightly rustling leaves, blue sky, no clouds, warm with great promise. Dad picked me up and threw me on the top of one of the girls, I don't recall which. I remember just how wide and massive her back was. My little legs stuck straight out sideways as I hung on for dear life to the Hames balls that were attached to the harness collar. I sat behind the collar and in front of the back strap. The smell of the leather mixed with the sweet aroma of the sweat under the back-strap pads added to the experience. There were also sounds: the jingle of the heel chains on the end of the tugs, hung on the rump strap until needed for pulling, the sound of the leather as it moved and squeaked and creaked with each movement under it.

I don't recall how long I rode, but it was a while—a very special time for a little boy. I believe that was the last major job the team was involved with. They were around and in the pasture for another year or two; however, they really had been replaced by tractors and were no longer needed. A man with the last name of James stopped by one day looking for heavy horses to make up a carload (rail boxcar) headed to Ontario. Reluctantly, the girls were sent on this adventure to eastern Canada. I'm sure my father missed them, but probably not as much as my mother, who always had a soft spot in her heart for horses. It was the end of an era, one that I am pleased that I had the chance to experience even as a very young boy.

# Chapter 5
## Spring Chickens

Little people and little animals seem to be attracted to each other. In this case it was little birds; baby chicks and turkeys. My older brother, Kent, and I had acquired important information, probably at breakfast, on the day in question. It was a sunny warm spring day, and we had received notice that our grandmother had received her baby chicks. I think we may have been told that we could see them but that Grandma wasn't home this morning so we would have to wait. I don't recall the discussion between two little boys, but it was decided that we had to see those chicks. We had been given white naval captain's hats by some admiring relative. These were no ordinary hats; they had pure white backs with a sleek black peak. At the front was a very majestic naval emblem, probably a crown and anchor, and they had gold braid, attached by a shiny black button at either end, laid impressively across the peak. So we placed these on our heads, having removed them from the moose antler rack hat hanger that Dad had built that graced one wall in the front porch and began our journey.

Our grandparents lived on a farm a mile and a half east of our farmstead. The grid road that ran from our place to theirs was a narrow gravel affair. It had been built as most grid roads of the day. A cut was made along the prairie with a four-foot perpendicular drop on the outside edge of the resulting ditch. This sloped up to the road shoulder, with the removed soil being shifted to the middle of the road surface to build it up. This design had its problems, including using black organic soil for the road surface (which wouldn't pack), and these roads were infamous because of the snow drifting that occurred in the winter. The edge of the ditch was several feet from the fence line. These early fences were quite often made using diamond willow as fence posts. This prairie willow had the property of not rotting quickly, and thus when used as posts would last for years. They were rather hard, especially given a chance to dry. Once hammered in, the staples were very hard to remove. Because diamond willow is twisted and knotted, these fences were very picturesque rather than "fence-line" perfect.

Along the fence line were small bluffs of wild fruit, such as saskatoon berries, chokecherries, and pin cherries. Between these and the ditch the ground was covered with prairie wool, made up of the fescues of the prairie and the other flora, like buckbrush, growing alongside it. In different spots along the ditch there were small poplar saplings, leftovers from the road-building days when the roots that were left in the soil gave rise to new growth that sprung up again. Often as you walked along in these clumps of bushes and trees a bird would fly up; and since it was spring, if you looked carefully you might find a bird's nest. We had been taught to carefully watch and look at these nests but never to touch, as the mother might not return if we did. On most walks along these ditches at this time of year at least one nest would be found.

Another interesting feature near our home in the south ditch was a crude oil or tar pit. This was in the ditch because of the mishap of a Ronaghan's oil tanker that had flipped over and dumped its contents there. Those were the days when the recovery of the truck was all that was done; the tar lay there for years. It sort of formed a light crust and was dusted over, but on real hot days it was soft enough to be a bit liquid. This quite often occurred when the sun shone down in the ditch on hot summers' days and the black surface of the tar absorbed the heat. I remember that if you ran real quick over it you could leave tracks without breaking the surface and getting the gooey guck on your feet. Naturally, of course, we on many occasions pushed the limits and came home with a tar mess on our feet to clean up. It did have its unfortunate side as well, as sometimes blackbirds and such would land on the surface and sit too long and suffer the consequences by not being able to get out. The crude oil spill remained there until the road was regraded many years later.

So off we went along this interesting road on our way to Grandma's house. I don't remember but its quite possible that our trusty dog of the era, Bugle, was by our side. That may have also given our folks another clue; two boys and a dog missing. At any rate, we made it. I still remember going into the old farmhouse, calling for Grandma, and wondering why she wasn't there. She was always there.

Grandpa and Grandma's home was of the type typical of that time on the prairies. You came into the porch, which had the porch-side chest-type cover of the wood box. This was a very important part of the entrance, because all the cooking and much of the heating in the kitchen was provided by the cookstove or range and these used wood for the fuel source. This outer porch also had hooks and places for boots that were shed before one entered the kitchen dining area proper. On the right as you entered was the wash-up sink with the hand pump that would bring up water from the cistern. Hot water for washing was brought from the reservoir on the kitchen stove and cool water was simply pumped into the washbasin. On the south wall stood this kitchen range with its reservoir, upper warming oven, black cooking surface, and element removal handle. To the east was the inside opening of the wood chest. The wood was placed in the outer porch part of the chest and retrieved by opening the inner chest door and placed in the stove's firebox. Thus, especially on very cold winter days, it was possible to get wood for the fire without having to go out to the unheated porch.

There was, of course, the inevitable kitchen table surrounded with chairs and the couch that was used for the after-dinner nap. These naps were what kept folks going who had been up harnessing horses at dawn and walked or ridden behind a dusty implement for fourteen hours or more a day. It was amazing what fifteen minutes on this couch could do to rejuvenate mind and muscle following the noon meal.

It was between the cookstove and the wood box chest that Grandma had fenced off an area, covered the floor with newspaper, and placed the baby chicks. The cookstove again was called on to add warmth to the area, and seemed to be doing its job, as the chicks were all lively and not bunched up but instead busying themselves around the feeders and waterers. How wonderful it was to see these little birds as they cheeped and pecked and generally ran and flapped around, seeming to tell us just how great it was to be alive. We watched in amazement and wonder. I don't remember picking any up, but it is possible that we did. We had been taught to be very gentle.

Well, all good things must come to an end and I remember the hum of Dad's old brown Dodge truck as it drove into the yard. He had tracked us down. I don't remember even getting scolded on the return trip; the chance to see Grandma's chicks probably has erased this from my memory. The chance to see one of the sure signs of spring has even erased the memory of the long walk; all I remember is the delight brought to a pair of small boys by their grandmother's new chicks.

# Chapter 6
## Teko

Teko Heilo Colantha was his full name. Teko for short. He was the herd sire that was housed on the farm during the early dairying days of our parents. I believe he had come from Ontario through our local vet, one of the Saville boys, Max, and was registered in the Holstein-Friesian herd book as having purebred parentage on both sides of his pedigree. My father was always a leader when it came to this sort of thing. When many were only interested in a "cow freshener" to keep the milk flowing, my dad, with encouragement from my mother, knew that heifers from a good herd sire would eventually pay off with long-living, high-producing cows. I know my mother was especially interested, as it was she that kept the paperwork of registration up and did the marking drawings or took the pictures of the calves and the general bookwork on the cattle. My father made sure that the cows were milked, fed, penned, and fenced properly. They were a great team.

Teko had his own corral. It was built in a circular configuration. Two posts were erected every fourteen feet or so with a gap of ten to twelve inches between them: one post on the outer circle, the second running inside the other. This continued all around the circumference of the circular outer edge of the enclosure. At the northwest portion of the corral was the entrance gate. It opened into a log alleyway that headed towards the slab barn. The gap between the outer and inner posts was wide enough to put a log rail in. These rails were placed in every other gap around the corral in an alternating fashion and the ends were reversed on each round. This was necessary to keep the rails even, as the butts of the rails were larger at the bottoms than the tops. These rails of poplar or balsam poplar were found in low-lying areas on the farm. They had no doubt been skidded from the bush to the corral location by Lexi and Flicka. Horses were very good log skidders and had been used in the bush camps where Dad worked in western Alberta in the early years of our folks' marriage. Keep in mind that these trees had all been felled and hewn by hand, and all the limbs removed by Dad's axe as well.

It's no wonder that Dad and the hired hand of the day, Howard Allenbrand, had no problem eating too much. The calories of a large breakfast of porridge, bacon and eggs, rich cream and toast, a huge noon meal, a coffee break with cakes or cookies, similar fare at supper, and a bedtime snack were easily dispensed with given the rigours of this routine. Makes today's gym routine seem rather pale by comparison.

So the corral went up in this fashion and each set of posts were tied together with number nine wire once all the rails were placed. This was important so that the whole structure didn't collapse. It was a very strong, solid enclosure with six-inch spaces between each log made by the ends resting on the log next to it. It was easy for us to view Teko without fear of him being able to get to us. Not that Teko had a mean streak and was out to get small boys, but, as with any animal his size, he was unaware of his destructive strength. He had an old truck tire and a pole in the middle of the pen to occupy his desire to paw the soil and push things around. More often than not a rail would be boosted off the top of the corral because Teko was scratching his horns. We often would load up our Radio Flyer wagon with what we thought was an inordinate amount of grass, dandelions, and other tasty flora to push through the gaps into his feed trough. Usually a couple of swipes with his massive tongue and a half hour of our work was long gone.

Two small boys in the forty-to-sixty-pound range enjoyed being scared to death by this large individual. I remember his eyes looking beady, but they were large. They just looked small in comparison to his massive head and rolling neck and shoulder muscles. He was quite handsome when you think about it: beautiful black and white coat in large separated patches over his body, a mostly black head, black sizeable horns, white legs and feet, and a white star in the middle of his forehead. As he walked you could see the muscles rippling under his shiny coat and his brush-like tail was forever busy, moving the flies along. Dad said even though he was not fat he weighted 2500 pounds when he was eventually sold.

We liked to watch as Teko ate, licked the blue salt block (which meant it was cobalt iodized), or ate his forage. The most fascinating part was how he managed the brass nose ring. All large bulls always had a ring in the nose for handling. This very sensitive part of the anatomy was used as the control point. As long as you had a hold on the bull ring, you were in charge of the bull. I've seen bulls walking sideways wanting to do their own thing, but the handler was able to keep them in check by keeping a firm grip on the nose ring. Teko's ring had a short six-inch piece of chain dangling from it. This was easily grabbed and extended with a snap chain when handling him was the order of the day. He was very adept at moving the ring up and down as he wiggled his enormous nose. He could make the ring sit right on top or, depending on nose movement, drop it to the bottom, where it dangled just above his mouth. Of course, because it was always moist, its being made from brass was very important. It didn't rust.

Teko was around for a few short years; I'm not exactly sure how many. As any good student of animal husbandry understands, inbreeding can present problems. His first heifer calves were soon old enough to become cows and could be bred around the eighteen-month to two-year mark. This was the time when the new phenomenon of artificial insemination was in its infancy. Again Dad was a first to be involved in this area. Live cooled bull semen could be shipped from Milner, B.C. and used to inseminate our cows. Not only did this eliminate the need to maintain a herd sire, but it gave the user the availability of genetically superior livestock to become the sire of the next group of heifers. It sped up quality gene selection a great deal. At first insemination was done by the veterinarian, but later it became a business for an English immigrate named John Davis, who lived at Edgerton. As I recall, he drove a Volkswagen Beetle and would come and inseminate your cow if he was phoned in the morning. During this time the handling and extending of bull semen and the methods of freezing and thawing made great strides in becoming available to prospective buyers across the prairie provinces—and the western world, for that matter. Many years later my education was in the animal science field and specifically animal reproduction. For many years I handled all the cow breeding on the farm, including the thawing of the bull semen and all the paperwork that was part of the process. It did raise eyebrows a time or two when the guests we were entertaining were uninformed about the process involved and my wife Aileen would say, "he'll be in shortly, he's out breeding a cow."

I don't recall when it happened but once the artificial breeding method for the dairy herd was well entrenched, Teko became expendable. I believe it was Mr. James, who ran a small trucking company, who came and loaded Teko for his ride leaving the farm. Although his time there was not long, I know his influence was in the background of our dairy cows for many years; if one was to follow the pedigrees for a number of generations, perhaps that influence was still there many years later. I do know one thing for sure: his memory will live forever and always be there. That big an animal, that small a boy; it is impossible to forget.

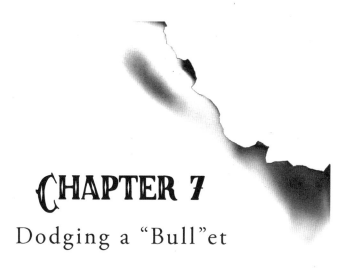

# Chapter 7
## Dodging a "Bull"et

Animals and birds of all descriptions take parenting seriously. As children we were forever going for walks in the back pastures, and I especially remember the many birds' nests we would find in the springtime. It wasn't uncommon to come across duck nests, for example. We had been taught to look but not touch, and to give wide berth once we knew where a nest was located, because there were always scavenger eyes watching and waiting for a tasty treat. Early in the nesting period ducks and other birds could leave the nest if disturbed. It was always interesting to watch the duck run under cover for a distance and then fly up, rather than fly right off the nesting area, so as not to give up its location. The parenting instinct became especially acute when the eggs were hatching or about to hatch. Then you were in for a treat as the mother feigned a broken wing, flipping and flapping just out of reach and attempting to lead you away. This was particularly true if she had her clutch with her and they were on a walk somewhere. Certain noises she made would send them all into hiding in a flash.

Another great faker was the killdeer. They needed to be topnotch, because their nest-building efforts were not what you would call pristine. A couple of small rocks, a few twigs or pieces of old straw, and that was it except perhaps for a small indentation in the soil. The eggs' colour did help a lot, however, because they were a greyish white, mostly grey, with a mottled splash of reddish brown all over the egg. They did blend in well. You could be standing right over the nest and still have trouble locating it. And of course the racket from the displaced pair was very persistent. As you walked towards them they immediately became "ill and injured," hoping they would be able to lead you away from the nest or their little ones.

So this instinct of protecting the next generation was very visible to us as we watched it in action so often in the wildlife on our farm. More interesting to me was that in our domesticated dairy cows the same instinct seemed to be intact. In most instances the dairy cows' offspring were with the mothers a very short time: hours or perhaps a part of the day was all and therefore maternal attachment seemed for the most part to be non-existent. But it was very much there.

I believe the big black cow's name was Rhonda. She was near to calving. Usually when in this state they were left near the barn, but in this case Rhonda obviously calved early. We kids were dispatched to scour the back forty to find the calf. We looked over prairie and in every bush, to no avail. Rhonda came in with the rest of the milking herd and returned to the pasture, but we could not locate her calf. We knew she should have one out there somewhere, but where? The second day Dad tied a bell on the old girl so that we could follow her more easily. Sure enough, just before sundown, out near a bush stood Rhonda with a beautiful black youngster. We knew we had seen it from a distance but by the time we got there it was nowhere to be seen. We combed the bush nearby and there, in a small patch of buckbrush, head and body flattened to the earth, was our calf. A motherly call from Rhonda and the calf was up and running. We managed to track it down and get it into the old half-ton for its ride home.

Although on most occasions our Holsteins were well-behaved, there were a few times that wasn't the case. After all, any cow that was a daily pain to deal with was usually given its "walking papers," as it were. However, when it came to separating cows from their calves, it always paid to be vigilant. On more than one occasion that deep-throated beller of panic from a calf was enough to turn the tamest, most docile mama into a force to be reckoned with. Of course a small boy or even a grown man was no match for the fury displayed by one of these mothers. I do recall one incident with a cow, Irene, I believe, whose actions sent me over the nearest fence in a hurry and left me figuring out a new strategy for getting her calf. This was when I was slightly older, just in case you thought my father had death wishes for his children.

This brings me to my most memorable occasion of animal parental protection. Teko had to be watered every day, as his corral did not have a water source. As you can imagine, this was a constant chore. In today's world it would have been taken care of with an automatic waterer. In those days the whole herd was watered from a water trough, with different pens coming up to a portion of the circular fixture. It had been made by nailing planking together and cutting it into a circular configuration for the bottom. As I recall, this was about eight to ten feet in diameter. Next, planks were cut in two-foot sections and laid side by side, nailed together by laying a piece of number nine wire near the top and bottom and using fencing staples to attach to each individual piece. The width of these pieces, when added together, needed to be long enough to be the same length as the circumference of the bottom of the tank. Next, all the pieces were painted with a slurry of copper sulphate (bluestone) to help preserve the wood from rot. Finally they were turned up on end and brought around the bottom to form the sides of the trough. An upper and lower steel rod of probably three-eighths-inch steel was wrapped around the outer circumference of the standing stakes and fastened with a connecting buckle that allowed you to tighten the rods by turning a washered nut onto the threaded end of the rod. This gave a firm tightening of the wooden pieces and brought the standing stakes tightly against the bottom and each other. At this point the tank was nearly waterproof but not quite. Of course, the swelling of the boards from being wet did help, but it was the addition of a pail or two of grain chop that filled the final cracks and made the trough hold water. This food source and the constant small grain chop droppings from the cattle as they drank made it an excellent place to have a few goldfish. Dad was great at doing things like that to make life more interesting. His motto was not "Why?" but "Why not?"

So the water trough was the main drinking area for the herd, with the water being supplied by a newly installed pump in a small pump house. Teko got thirsty every day. The day in question I was out helping as only a four-year-old can. I had probably been placed next to a fence and told to stay there while Dad watered Teko. As it happened that day, one of the cows had calved and she and her calf were wandering around in the herd of cows a short distance from the watering trough. Dad decided to multi-task that day and leave Teko drinking water while he went and removed the youngster and took it to the calf area in the barn. Now, understand that Teko did enjoy his time at the trough. He savoured every moment, pausing and then drinking and pausing and drinking some more; keep in mind he wouldn't be drinking for another twenty-four hours. In the meantime I had ventured out from the area of the fence and had entered a sort of spillway area where the water trough had overflowed and left a muck-hole down through the middle of the pen. I slowly entered the quagmire and was about halfway across in my little rubber boots when Dad got ahold of the calf. The calf decided right then and there that it didn't want to be taken from its mother and let out the distress call beller. The cow herd quickly began converging on the situation with bawls of their own and Teko raised his large head, pricked up his ears, turned on a dime, and began towards the cowherd at a trot. There I was between Teko and the cowherd as they quickly converged. What could I do? When one is under duress it's amazing what can transpire. In a flash I was out of my boots, running for cover in my stocking feet. I crawled under the fence and made it out of the pen as quickly as my little legs would carry me. I don't recall what happened to the boots, or how my mother received my manure-covered socks, but I did learn one thing that day: fear is a great motivator.

# Chapter 8

## The Hog Lot

I have faint memories of the hog lot, one of the early ventures on the farm. It was located south of the present log barn and was always an interesting setup, especially since it seems the keeping of livestock has in some ways made a full circle. These were definitely "free range" swine. The lot consisted of a fenced-in area that included a fence line of posts and wire: barbed wire on the top and bottom with hog wire in between. Hogs love to root around with their noses and thus it was important to have a barbed wire at or just below ground level in order to keep them from rooting under the hog mesh and escaping.

One side of the lot had a plank fence with a trough for feeding the herd. There was also a bored water well right next to the trough. The water had to be raised by a hand pump and directed into the trough to keep the thirsty mob hydrated. The well was bored by a gentleman named Cecil Ricker, I believe; he had a well-boring machine. It seems to me he had something to do with the flour mill in Wainwright as well; however, these facts are not well remembered by a boy that would have been just a few years old at the time. I do know that two wells were bored by him on the farm: one at the hog lot and one near the house. These had casings of wood dropped into them and, I believe, between the walls of soil and the walls of the plank casing concrete was mixed and introduced. The finished wells were about twenty-four to thirty-six inches across, and they were covered with wooden planking covers. I remember the well at the house had a dipper hanging from it to get a drink any time you were thirsty. That was, of course, after you had pumped the water with the hand pump and rushed around in time to catch some water in it. As far as I know, the old pump at the farmhouse still works.

The housing for the herd was made near the end of the threshing machine era on our farm (when all the crop was put up in bundles and hauled to the threshing machine to be processed). The thresher was set up near the hog lot with the straw discharge pipe pointed high and into the lot. The bundles and straw were the culmination of a summer's work, starting with seeding the land. The crops had the long summer days to grow and mature quickly. The next step was the cutting of the nearly mature crops. This was done by a machine called a binder. For many years this had been drawn through the fields by horses, but by my time the horses had been replaced by an 8N Ford tractor owned by my grandfather. The machine cut and tied the grain for handling purposes and required an operator sitting at its rearward side to lower and lift the table, drop the bundle collector, adjust the reel, and so on. The slowly rotating reel pushed the standing grain across the knife or sickle. A rotating canvass moved the cut grain to an elevating upper and lower dual canvass at the end of which the material was taken up a steep incline and let slide down a forty-five-degree decline into the knotting area. Here the cut grain bunched up until enough had accumulated to make a bundle of material. It then triggered a reaction whereby the bundle was tied tight with twine, the twine was knotted, and the bundle was kicked out onto a carrier. When eight or ten bundles had accumulated, a foot lever kicked by the operator dropped the bundles in a pile and the process started all over again.

Next, workers on foot came to each dumped group of bundles and stooked them. This involved picking each bundle up and setting it and its companions on their butt ends with the heads pointed upward. Here the grain was left to finish maturing and to prepare it for threshing. The stooks of bundles were then loaded by hand into a bundle wagon using a pitchfork and then taken to the thresher. Here the bundles were pitched onto the large-mouth conveyor of the thresher and they entered the bowels of the trembling giant to begin the process of separating the grain on the heads of the bundle from the rest of the materials.

This stationary machine was what was sitting backed up to the hog lot waiting to process the bundles that were brought to it. I'm sure that Lexi and Flicka would have been involved hauling the bundle wagon from the field to the threshing machine. They were required to walk up alongside the racket of the whirling machine with its vibration and multiple moving belts and pulleys without incident. This was common across the prairies for many years and I'm sure old horsemen could relate many tales of the way horses handled this situation.

In later years the threshing of crops took the machine to the crops and not the crops to the machine. In other words, the cutting and threshing of the crop were combined. Thus the term "combine" was born and now is synonymous with large crop-processing machines that traverse the world's croplands separating the grains from the remaining residue.

The straw from the threshed bundles was blown over a clump of diamond willow trees and thus left the stack with hollow areas under the covered spreading branches. The grain would have been conveyed to the wagon by an elevator in the threshing machine and would have slid down a discharge pipe into the awaiting wagon box. This grain was a very important commodity, as it would make up the majority of the hog feed for the winter. The insulating value of the straw, both for coolness in summer and warmth in winter, was one of the greatest features of using the crop refuge for this type of housing. The bedding value of the straw was excellent because pigs are intelligent and do not mess in their bedding but do their business in a corner away from that area. It was a great place for sows to raise their broods and come out in the spring with a batch of little ones that, when grown, added to the farm income.

And so it was that I was present when the threshing was going on and the straw was being blown over top of the diamond willow. The power to run the equipment was being provided by the 15-30 McCormick as it was connected by the long belt drive to the threshing machine. I was out "helping" and I distinctly remember that I was cold. Threshing naturally took place in the fall when the grain crops were ready to harvest. There weren't any leaves on the trees and the surrounding countryside had that ready-for-winter look, so this threshing was taking place when cool weather could be expected. It was not uncommon to do threshing in the late or very late fall or even sometimes when snow had arrived. The stooked bundles remained rather dry even in inclement weather, such as can happen at this time of year.

Not being involved physically meant that, standing and watching, one could get cold on days like this, and so my father brought me up beside the tractor and had me stand near where the large fan was drawing air through the radiator. I distinctly recall it was very warm there, with heat being transferred to the blast of air flowing around the engine. I also remember his warning that it was okay to hold my hands up to enjoy the heat but not to put my hands too close to the fan because of the grave danger if one caught their hands in the blades. I certainly wasn't outside with the men for the whole event, but I'm sure I was there long enough to give my mother time to catch up on some things in the home while my younger sister slept. As to the hog lot, there was a decision made sometime in those early years to concentrate on the dairy part of the farm operation. We never had hogs on a large scale again. We did, however, usually have a couple on hand to eat the scraps from the peeled vegetables from our household, to root around their pen, and to supply the bacon and pork chops we all enjoyed so much.

# CHAPTER 9
## Good Neighbours and Friends

Farming has a huge advantage when it comes to children. Many times in our childhood we were at our parents' sides, helping and learning from them as they worked. It's hard to imagine, but this isn't possible for most children; they don't get this chance. A surgeon's child, for example, cannot accompany him/her into the operating theatre. In our case, however, it was second nature for us to be with our parents whenever they were working around the farm. And we weren't just observers. Children did all kinds of little chores, from feeding calves to collecting eggs, running errands, and helping in the garden.

In the spring it was very important to get the garden planted. Sometimes Dad would hook up the two-bottom plow on the 8N Ford and would go to the garden at potato planting time. He would plow the first furrow and Mom and us children, sometimes along with my maternal grandmother and others, would fill the resulting trench with cut potatoes. These plantings were carefully prepared to have a couple of potato eyes per piece in order to have a root and stem starting point. Of course, one didn't want to be wasteful by using whole potatoes, as there was still the time between planting and harvest to feed the family. Once the trench was planted, another pass with the plow covered the first group of spuds and opened up the next row for planting. This continued until several rows of potatoes were in the ground. After all, a hungry family needed a lot of these pleasant starch sources to carry them through the year. There was nothing like mid-July fresh potatoes with my mother's gravy. I can taste it now.

Children could also be taught to plant peas and beans without much problem. Their little hands could easily place these large seeds in their proper place in the rows, and being close to the ground was also an advantage. When it came to planting the more delicate seeds, this was ruled a more adult type of proceeding. I do remember that our grandparents had a garden seeder that had a small hoe-like soil opener, two small closers, and a dual packing wheel behind. The planter had a lead wheel and handles towards the back, which were used to push it along. In the centre circumference of this lead wheel was a gear that drove a small mechanism attached to the seed box that metered out the seed. A graduated slide on the side of the box helped you set the seeding rates for various garden seeds: carrots, radishes, onions, corn, you name it. You could always tell when it was operating as it made a constant clicking sound. It also had a metal arm attached to the back of the seed hopper that could be swung to either side. This had a piece of chain dangling from it and when this was lowered to the side it would mark where the next row should be. Depending on how well the first row was lined up, all the next rows would follow suit. It was important in our household that things be done well, so the rows had to be straight.

Many years later there was the mishap of misplaced potatoes. A retired, single railroader used to come out to the farm to hang around and help out where he could. He also enjoyed the company and I'm sure the home cooking. At any rate, he was relegated to help plant potatoes with my no-nonsense grandmother. Each stood at an end of the rows and were to be moving the stake for the next row upon completion of the present column. All seemed well until the potatoes emerged. There in the garden were two rows of potatoes with row width at one end, converging on the other in a perfect V. Nobody would admit to not moving the stake, and it was a wonderful conversation piece.

We would get Grandma and Alf debating the missed stake move, and duck out while the debate was being hotly contested. Made for a lot of fun as no one would admit to it. We called them the "V for Victory" potatoes.

Of course, the field crops were being planted at the same time as the garden. In the early years oats and barley were used, as they were great when cut and used as green feed and, in later years, silage. It was the grain crop that was allowed to mature for the seed that was threshed from the straw. This grain was then hammered into chop when run through a hammer mill powered by the belt drive of the old McCormick. An old high granary on skids stood next to the barn. It was divided in half, with the divider boarded up four or five feet high. In the back, half the ground-up grain was blown from the hammer mill. I'm sure this was a back-breaking day for Dad and perhaps the hired man as grain was shovelled through the hammer mill to fill the chop bin. It doesn't sound like fun getting a tractor going, shovelling on and off a wagonload or two of grain, if it was minus thirty degrees Celsius outdoors, often with a breeze blowing. This lasted for a few weeks and all livestock that required chopped grain would be fed from there. I do remember that chopping day was centred around a break in the winter weather, if possible.

I didn't have too much to do with feeding at that time, but I do recall the fun of catching mice that ventured into the bin. You see, it was easy to cause an avalanche of chop with the scoop shovel in hand. The chop covered the mouse and then it was our job to get it by the tail before it could dig itself out and take it to the cats that were constantly hanging around. It makes one wonder if the cats of today carry some of the genes of these early tabbies; a lot of generations of cats have come and gone, I know.

While our family was busy with spring planting and all the activities that were required to make a success on the farm, so it was with the neighbours. Because of all this busyness sometimes one of us children would disappear. It must have been a constant worry for our parents during this time with the first three of our family all running around the farm. It was during one of these quiet times that my younger sister, Alison, decided it was time to take her doll for a stroll to visit a friend. She was a happy, determined young miss, hummed tunes a lot, without a care in the world. And so she started out pushing her buggy before her and walking up our east road. Fortunately one of our neighbours happened to be working in the next field and actually changed his working pattern so he could keep an eye on this wee one. He worked along the fence line to be sure that she didn't come to any harm and was there when my folks came looking for her. It was great how the neighbours looked after one another. The Smedberg brothers had come from Norway to farm in western Canada. Sverre and Fred were the brothers' names, with Fred being the married one. As I recall, they were fine neighbours; my mother got to know Mrs. Smedberg through the local ladies group of the Sydenham-Gerald FUA. So Alison was returned without incident and was no doubt made aware that she wasn't old enough yet to strike out on her own.

As mentioned, the children were an important part of the farming operation and the errands they undertook saved many steps and helped out a great deal. On one occasion my younger sister Bethany was sent out to tell Dad that supper was ready. As a little tyke, she assumed that her dad was in the field and so she walked out into the field beyond the barnyard to find her dad and any others to tell them that supper was ready. The men were not in the field; they were probably in the barn or around the buildings in the yard. The crops had grown and matured since spring and were quite a bit taller than my little sister. Once the men arrived for supper, it was discovered that she was nowhere to be seen and the search for her began. All available farm personnel were thrown into the search effort, kids and adults alike. As I recall, we really didn't know where to start. I remember being out in the field calling and searching without success. I'm sure my folks, especially my mother, were beside themselves. After a very long search, I believe it was my father who saw the grain moving and discovered it was Bethany walking, not knowing where she was and totally disoriented. I'm sure there was great relief around the table that evening as we finally had supper; in fact, it probably was like another Thanksgiving meal. It probably tasted even better as we all rejoiced at the return of my little sister in fine form and without injury or incident.

# CHAPTER 10
## The Fine Art of Fencing

Quite often in the spring an inventory of the need for repair of the fences on the farm was carried out, to find the weak spots before the cows did. More often than not this happened in the reverse order: the cows found the weak spots and the repairs were made after the fact. I always thought that one cow was designated to go around at night to see if there was a spot for the herd to escape, but I have no proof. One thing for sure, if one cow was out, in short order they were all out. There is a saying that "the grass looks greener on the other side of the fence." It seems to me that a group of dairy cows could have authored that, for as soon as a cow was feeling comfortably full her nose would invariable go in the air and greener pastures were her mind's sole occupation, especially if they were on the other side of the fence.

Major fence work started with the gathering of a large number of three-to-four-inch poplar saplings that were in their spring mode of sending sap and nutrients up into the branches. These were cut and immediately set upright in a tank of water containing copper sulphate (bluestone). As the trees or posts drew up the water they also drew up the bluestone. This then made them much more resistant to rot and into fence posts that would last a long time. This happened fairly quickly: a few days in the water/bluestone mix and you could see the blue colouring under the bark at the top of the cut post. Another often-used method was to cut the posts to length, six or seven feet, and then peel the bark from the bottom two feet and peel one strip all the way up the post. This peeling was done with a drawing knife: a two-handled device with a blade in the middle and a handle on each side. You sat on the post and drew the knife towards you, peeling bark as you went. This continued until you had the bottom two feet and a strip up the entire post cleared of its bark. As the post dried the bark would come away because of this peeled strip.

My father had rigged a saw of sorts that had two blocks of wood covered with thin metal around their tops, fronts, and bottoms; these were bolted to the side of the tractor's front end. A shaft with a pulley on one end and a three-foot saw blade on the other was bolted to the wood beams with a pair of grease-able pillow blocks holding it in place. A belt ran from the pulley on the saw to the drive pulley on the tractor to provide power for the saw. The saw was used to cut the posts to length and then Dad would take each one and run the larger peeled end into the saw to sharpen it. It sort of looked like a pencil when it was finished and, of course, was much easier to drive into the ground when sharpened. If a new section of fence was being built a sizeable number of these posts were made ahead of time to be ready to build it.

Dad loved to have a good-looking fence. It needed to be straight with all posts in a line and the wire needed to be at the right tension with exactly the right width between strands. In our case most of the barbed wire was recycled: rolled up from the previous fence or some other fence recovery. I do recall the differences between the barbed wire. It was easy to identify different types, especially those made during the Second World War. The wire was a much lighter gauge, thus saving the precious steel that was needed in the war effort.

The fence was started by digging in a strong corner post. This was usually dug in with a post-hole auger and shovel. The post was placed in this post hole and tamped in using a tamper made from a piece of light pipe or rod about six or seven feet in length with a small chunk of metal or large bolt attached to the bottom. The top sometimes had a blade-like piece of metal attached to it. When turned end for end, it could be used to shave the post hole in order to accommodate a post larger than the diameter of the post-hole auger. It was also used to cut small roots that were within the hole circumference when you were in a treed area. So the corner post was set by dropping it in the hole and using the tamper and the removed dirt to set it. It must be straight up and down when viewed from all directions, and tamped with the tamper by pushing some dirt in and tamping with an up-and-down motion to make the post very solid. This continued until most of the dug-out dirt was tamped in the hole around the post and the post was immoveable. Next the whole process was repeated about eight feet away with a second post and a wooden rail was placed horizontally and nailed between these posts about six to eight inches below the tops. A strand of wire was run from the bottom of the corner post to near the top of the companion post and back again on the opposite side, forming a loop. This was then tied together and stapled at the bottom of the corner post and the top of the companion post. Next the two wires were wound together using a rod or a piece of tree branch, anything to wind them tight. The resulting corner made for a very strong braced structure, able to be used to tighten the strands of wire that would make up the fence. This whole process was repeated at the other end of the length of fence being built. Sometimes if the fence was over a half mile long an additional set of posts were erected in the middle of the span in order to have an anchor for tightening. In this case a loop of wire was run both ways and twisted so that the twin posts were both braced to accomplish this tightening in either direction.

Next a roll of wire was unwound between the two fence corner posts and tied at one end in the correct position, usually starting with the top strand of wire, to avoid entanglement of wires along the fence line, and then hooked onto the wire stretcher at the other. Our wire stretcher was made with a rope running through a series of pulleys that provided a mechanical advantage when tightening the wire. Two pulleys and a wire clasp were on one end and two pulleys and a rope lock as well as an anchor chain were on the other. The wire was placed in the wire clasp and the anchor chain wrapped around the corner post. By pulling on the rope the fence wire was stretched between the corner posts at either end of the fence. If you lifted the wire and whipped it side to side and up and down, it settled nice and straight between the corner posts.

This is where the real work began. My father would walk along the wire, stepping off four or five paces between posts. A crowbar was used to start a hole where the fence posts would be placed. This was a five- to five-and-a-half-foot bar of steel with a tapered shaft and a triangular end that was used to drive into the soil to make a hole. Naturally, in the spring the soil was usually moist and so it was easy to make the pilot hole; however, I've seen it when it was very dry and the hole was started by needing to add water to each hole. Sometimes Dad returned to add more water a number of times and then enlarge the pilot hole to accommodate the posts. Next, the posts were placed standing in the pilot holes and then pounded in so that they were in the soil a ways, were solid, straight, and true. This was all done from the accompanying wagon holding the fence supplies, or perhaps a used five-gallon pail that served as a step. After a day of swinging a ten-pound maul hammering in posts, a good meal and comfortable bed were a welcome sight.

This driving posts did require technique as well. It required a good eye to be sure the mallet hit the four-inch post top. It needed to be administered so that the mallet head met the post top square and true. You could feel the "sweet spot" of post pounding when you got it right. There was a satisfaction and a solid ring when steel met wood correctly; there was an equally unwelcome sensation of vibrations in the hands that told the pounder when the swing wasn't ideal. I've personally taken a big swing only to miss the post nearly completely. As the weight of the mallet carried me headfirst over the side of the wagon and onto the awaiting ground and shrubbery, a lesson on post pounding accuracy quickly gained importance. Naturally this lesson took place when I was somewhat older and able to swing a post pounder, but it was a good lesson just the same. Another lesson was taught to us by our father. As the posts were pounded, turns were usually taken by a couple of men. One person held and straightened the post and the other pounded. Dad told us the story of a neighbour he had that was pounding posts with a young guy named Hadily. It seems Hadily had a fair amount of brawn but lacked the same strength between his ears. He placed his hand on top of the post to wiggle the post and straighten it while his companion brought down the mall with great force for the next drive. Hadily only had nine fingers after that incident.

Interestingly, in later years when I was somewhat older, the midways at Stampedes and fairs had a game where you paid your dues and were given a large mallet to hit a pad on a fulcrum, which sent a piece of steel flying up a rail that hit a bell. If you could ring the bell you got a prize of some inferior stuffed toy: a great gift for a girlfriend. Needless to say the operators hated it when the farm boys that had been building fence showed up. They had the technique and the strength to ring the bell every time.

And so the post pounding continued until all the posts were in the ground and they were lined up with the first strand of wire. There was a certain satisfaction to standing behind the first post and not being able to see any others because they were so straight and true that none were out of line. Next, the stretched wire was raised to the predetermined height and stapled. This height was adjusted depending on whether this was a three- or four-wire fence. The staples as well had to be carefully driven. They were not to be right tight but should have a very small amount of play so that there was a bit of give to the wire when animals leaned on the fence or poked their heads through it. If this give wasn't there the staples or wire might fail and a hole in the fence would be the result. So each wire was stretched and stapled in turn until all were in place and a new fence was born.

Spring fencing had its challenges. It also had its rewards. The smell of the trees, grass, wild fruit, and so on in spring was unforgettable. Quite often the buckbrush along the fence line yielded other surprises. Many were the song sparrow, duck, woodpecker, and other small birds' nests found. I particularly remember finding a rough grouse nest once with fourteen eggs in it. Dad was very good at pointing out different interesting things as we went along. It could be different rocks and how they were formed, old trees that had very crooked trunks in places, probably the result of Natives marking a path by bending and twisting saplings to be able to follow later, or perhaps a hawk diving to pick up a mouse.

One of my earliest memories, on my grandpa and grandma's farm, was the time I was along with Grandpa Dixon and Dad as they repaired a fence east of the old farmstead. They were doing many of the actions previously described to get the job done. They were using an old hayrack to hold the fencing supplies, pulled by Grandpa's 8N Ford. Sometime during that fencing excursion, Grandpa decided that I needed to drive the tractor with him. What a day! What excitement! I was beside myself with the knowledge that I was going to drive; I was probably only four years old but perched on the seat in front of Grandpa like I knew what I was doing. It was a few years later when I realized that in fact there was a lot more to driving than what I had done that day (I had been doing a little steering). That said, I will never forget the kindness shown me by my grandfather as he made a little boy's dream come true.

# CHAPTER 11
## The Blizzard

The snow had been falling since early morning. Great amounts of it. The temperature had been dropping as well. This part was not unusual. Highs and lows on the prairies quite often change places and the resulting "combativeness" of changing air temperatures can deliver some nasty winter weather. The previous weather patterns, air temperatures, and general resulting daily weather are no longer in my consciousness; the resulting winter storm most definitely is. I was too young for school but my older brother was not. He attended school in Wainwright, as we lived on the south side of the road that divided school districts. We were what was known as Wainwright rural school children whereas across the road the kids my brother's age would have started school in the little rural school known as Sydenham. This practice soon ended as it was the era of the closing of these small one or two-room schools, and the busing of rural children to larger centres had begun. I believe Kent would have only made it to grade two before Sydenham closed.

At any rate, as the wind picked up and the temperature dropped we knew we were in for a real doozy. As we looked out our windows it was very hard to see; the driving snow and the low overcast blocked most of our sight lines, including the light. There began to be some worry on the part of my parents and my grandmother as the day began to close in and there was the realization that my brother and grandfather were both expecting to come home at the end of the day. The storm worsened. The school board responded too late; the schools should have been closed much sooner than they were and the children sent home. By the time the children and buses were released the surrounding roads were already blocked with drifting snow and the urgent matter of finding accommodation for the rural children began. Fortunately our Aunt Grace and Uncle Felix Currier lived in town and my mother was able to contact them about keeping Kent for the night. How he got to their place I don't know—perhaps he recalls; I do know that it was taken care of and he was safe.

The issue of informing my grandfather not to try to come home in this blizzard became paramount. Again the communication method was to try to phone Grandpa and have him stay in town, perhaps with the Curriers as well. Grandpa had not been well and had left his farm a year before. It was thought that perhaps the stress of farming was getting to him and he needed to do something that required less physical labour. Grandpa had gone back to his roots as an accountant and worked at the Canadian Forces Base. He had taken this job having had the experience of being the chief bookkeeper for Foss Tug and Boat Company in Tacoma, Washington, in his early life. (It was later determined that in fact Grandpa had abdominal cancer, which would claim his life in less than three years.)

By the time it came time to deal with Grandpa and inform him of the terrible blizzard raging outside and not to come home, the phone no longer worked. The howling wind had probably wound up wires or covered the poles and wires completely with drifted snow. It certainly was apparent once the storm was over that many of the poles and wires were covered. Phones those days were party lines, with aboveground wires connected with cross-arms and insulators. They were often subject to damage from lightning, windstorms, hail, and many other mishaps that left them dead. That's exactly what happened on this occasion. Try as my parents might, it was impossible to reach the telephone operator in town who would have connected them to the office where my grandfather worked. (At that time, all phoning was done through an operator.)

I'm sure a discussion was held and it was decided that my mom and dad would bundle up and try to drive the old blue 1952 Chevy to our neighbours, the Smedbergs, who lived a mile and a half closer to town, with the hopes that their phones were not down and they would be able to relay the message not to attempt to come home to my grandpa. The chores of milking and feeding were hurriedly finished up and all was made ready for the trip. I'm sure there was an important discussion prior to my parents leaving. Grandma would have been summoned from her home across the yard to come look after me along with my younger sister, Alison, and baby sister, Bethany. This was not just a quick "throw on your shawl" type of outing for Grandma. With Grandpa away, it would require a "banking" of the old T. Eaton Co. furnace with a few choice lumps of coal to keep some heat in the old house while Grandma babysat us three. Not that it would have been a catastrophe if things got below freezing, as neither house had running water yet.

My folks left. I recollect the description of their journey as Mom told it years later. Dad was the driver and Mom was tasked with watching the edge of the road for the grass that grew there. It was the only thing they could see, or should I say Mom could see, as through Dad's windshield everything was frosted up and he virtually couldn't see anything. This combination worked well for over a mile of the journey; Mom instructed Dad how to stay out of the ditch. I think it involved the window being cracked open once in a while to be sure if the edge was in sight. Unfortunately, at one point Mom hollered that they were close to the ditch but Dad thought it was the opposite ditch and the truck plunged into the deep fresh snow off the road. This was long before four-wheel drive or posi-traction was standard on farm trucks. The only thing left to do was to get out of the vehicle and walk the rest of the way to the neighbours'. Getting out of the truck was a task in itself. The wind was so strong it was nearly impossible to push the door open. Both of my folks had to leave through the passenger door as the driver door was pushed up against the snowbank. The walk towards the neighbour's house begin with the full force of the storm pushing and tugging at the pair as they fought their way along. The relief was great when their walking finally brought them behind the trees near the neighbours and finally to the Smedbergs' front door. They were graciously received, given warm drinks and food as well as sleeping accommodations for the night. No one was going anywhere in that storm.

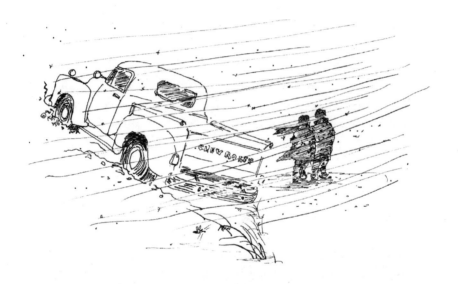

Back at our farm I can only imagine the angst of my grandmother, with three little children and no communication from her husband, daughter, or son-in-law. All there would have been was the worry of what had happened to all those she loved. The costs of being caught out in a prairie blizzard were familiar. Although not common, it was not unheard-of to perish in such a storm. I'm sure a number of prayers went up that night.

Grandma wouldn't have been without things to do. It was important to keep the kitchen stove going, again with the use of a lump of coal.

I'm sure my father would have filled the hopper of the Booker furnace with oiled stoker coal, so it probably didn't give her any grief. I no doubt went to sleep well-fed and having been read to by Grandma.

The next morning that ragging blizzard had subsided somewhat. Being a long-time farm wife, Grandma realized that the cows and other livestock required feeding and so I do recall us being in the warm moist barn, feeding the green feed to the cows. I suppose Grandma had brought us all with her and set up a place for baby Beth to be safe and warm while the chores were done. We didn't do the milking, as I recall. Dad did have a couple of old Massey Harris milkers but I'm quite sure Grandma didn't know how to run them. We definitely gave the cows their feed; I remember breaking open the bundles of green feed and distributing them to the cows. Grandma had been a schoolteacher and had an uncanny ability to get the best out of children no matter the circumstance. She was very positive, as I recall, and even in the difficult circumstances of the day never let on to us that this was not normal. We were rallied to do the job at hand like a well-trained regiment. The smell of the fresh oats and the green feed was refreshing and when mixed with the warmth of the moist air produced by the cows left a very memorable aroma. To this day any odour like this brings me back to my childhood and this venture in particular.

When all that could be done was accomplished, we returned to the house. The storm, although still with us, had subsided somewhat. I know not too far into the morning, my father arrived, having walked the mile and a bit back to our farm. I'm sure Grandma was delighted to see him and get the news of the safety of Kent and Grandpa. I'm assuming that Dad then milked the cows, put the strained milk in cans, and finished all the necessary chores before he hooked up Flicka to the stoneboat to retrieve Mom. I'm sure they were back in our own home by mid-day. I know that Dad had to haul the milk by stone-boat to the nearest place that could be reached by the retailers/processors.

It was some time before the blue Chevy could be retrieved, as the snow and snowbanks were everywhere. This was the year where Caterpillars with wide tracks were used to break through the snow to try to clear roads. In many cases the roads were abandoned and temporary ones were plowed out in the fields where the snow wasn't as deep and the trees and shrubs weren't available to catch the drifting snow.

The back-slopping of roads, the removal of these trees, and especially improved farming methods have all added to a much less hostile environment when it comes to winter storms. The practice of summer-fallowing land: the act of killing weeds for a season and keeping the land cultivated and black with no noticeable vegetation, has also been abandoned. During storms like the 1955 blizzard, drifting soil and snow mixed together made an absolutely rock-solid mixture that, as was mentioned, was almost impenetrable until the crystal structure was broken down by the warmth of spring.

It seems like every winter at least one blizzard-like storm comes to our area but the blizzard of 1955 will be forever etched in my mind as the one that left a lasting impression and deep respect as to the power of Mother Nature and how vulnerable we are to her whims.

## CHAPTER 12

### The Ditch Pasture

Weather had such an effect on many of the things we did on the farm. We were always looking to the west for signs of rain, it seemed. As children we knew the types of clouds, whether they be cirrus, cumulus, or cumulonimbus, and what a storm looked like when it was about to release its fury on us. I remember being in my grandparents' kitchen learning the directions—north, south, east, and west—long before going to school. These were important to know, in order to predict the direction of oncoming weather. I recall the early years of my childhood were years of very cold winters and hot dry summers. I'm sure this could be confirmed with climatic records of the area during these years. Severe thunderstorms occurred in mid-summer with very hot days preceding them. At times this fury would include the updraft of storm winds that would carry the moisture up and down several times, forming hail that would flatten everything in its path. There were the years where we received heat without much moisture and I certainly recall my parents' concern with dry pastures and small crops. Many were the times they prayed for rain to bring the crops on.

In the early years of the dairy farm the cows were fed in dry lots only during the winter. Other than that they were pastured south and west of the building site on pastures that were divided with fences; the cows were rotated in these different paddocks. When we had rain and the grass was growing well it was so pleasant to go out into the pasture and see the beautiful black and white cows against the various hues of green made up of the pasture grasses and legumes silhouetted by the accompanying shrubbery and trees. The smell of the cattle and the accompanying cowbirds that fed on the various insects surrounding the herd made for a perfect picture.

One of my jobs, especially in the summer, was to go out and get the cows for milking. By the time the afternoon milking came along the cows were usually full of forage and would be lying down chewing their cuds. They would lie in a cool area if possible, often in the shade of trees along the fence line or in a lower area of the pasture where the fescues were cool and thick. Their constant grazing of these areas made them just like a lawn and at times it was fun to lie on your back and watch the clouds go by, all kinds of different shapes and configurations appearing. Most anything could be imagined in the clouds. One couldn't lie too long, however, as the cows needed to be rustled from their comfortable beds in order to head home for the evening milking.

Each herd has a boss cow and a pecking order. One cow (the boss cow) would rise and all the others would follow. Soon the leader would head out on the nearest cow path toward home and the barn. If you are in a pasture you will notice that as you head to the yard there are pathways that the cattle follow. They start at the far end of the pasture and look almost like the tributaries of a river as they come together to a well-worn cow path nearer the yard. These pathways meandered around rocks, bushes, and other outcroppings in the pastures and always reminded me of a railway track with curves and straightaways. Once all the cows were on the path they indeed looked like a train with railcars as they slowly ambled towards home and the barn. Quite often I would grab the tail of the final cow and pretend I was the caboose and walk along behind her all the way home.

In these early years during the summer months, we children very often didn't wear footwear. Most often we could be found in bare feet. When getting the cows for milking I quite often found that stepping on prickles, small sticks, and stones could be quite painful. There was a solution that we developed to help alleviate this problem. As the cows rose from their beds, one of the first orders of the day for the herd was to defecate prior to heading off down the path. It was this fresh "cow pie" that was used to dip our feet in and then immediately into the loose soil of the pathway. This gave us a sort of sole for our feet to keep us from the brambles and the good thing was that if this one wore out the ingredients were at hand to freshen up the process. Needless to say, at the end of the day our final destination before we came to the house was the pump, where we pumped enough water into a pail to remove the makeshift footwear.

As small children we were not required to milk the cows. Dad used to milk the herd with a couple of old Massey Harris milkers. A small blue vacuum pump was housed in a lean-to in the northwest corner of the barn along with some crockpot-type vessels that held the solutions that were used to clean and sanitize the teat cups and milk lines between milkings. These, along with a sink and counter to hold milk pails and related equipment, pretty much used all the space. This "modern" way of milking came about because power came to our farm through the Sydenham/Gerald REA (Rural Electrification Association) in 1952. I would have been a strong two years old at the time and do not remember being without power. I do remember the little electric-motor-run vacuum pump but maybe my sister Alison remembers it better. For it was her beautiful curls one evening that were in danger when she backed her cute little head into the pulley and belts of the contrivance. I do not remember whether there was any missing hair from the ordeal, but there were tears and I'm sure a bit of an interruption in the milking that night as the wound-up hair brought the little pump to an abrupt halt. I believe both my folks were involved in the unwinding of the pretty locks.

Things ran pretty smoothly during the summer. Everyone was busy, my older brother was home from school, and we had responsible jobs at the level that we could be trusted to carry out. Dad and the hired help would seed the crops, put up the hay, and continue the everyday tasks like milking, feeding, cattle work, fencing, and so on. My mother would be in charge of the gardening and, with help from my grandmother, would make sure that the raspberries were picked, the wild saskatoons were canned, the garden was weeded, the Monday washday organized, the children fed, and the clothes mended. There was never a dull moment. This was all done at times while my mother was pregnant with a couple more younger sisters during these years. Life was not easy; there wasn't time to be sick or complain. With the years of good rainfall and good crops the summer went by rather peacefully, but there were many "character-building" years as well.

When the years of insufficient rainfall were part of our lot, there had to be changes to the routine. Well do I remember the years when the rainfall was insufficient and the pastures began to fail. This, of course, led to poor milk production, the all-important ingredient that was keeping the whole outfit afloat. This is when my mother, a registered nurse, would return to nursing in the Wainwright hospital in order to keep enough money coming in to pay the bills and this is when all available bodies were put into service to keep the farm alive. As the pastures declined there needed to be a solution to the feeding of the dairy cows and that's when the mile of ditch in front of the farm became the third pasture.

And it made perfect sense. All along the ditches between the road and the fence were many succulent patches of forage. From the fence to the four-foot ditch cut were the remains of the fescues and other grasses that made up the "prairie wool" bouquet that had covered the plains and provided sustenance for its occupants for eons. In the disturbed portions from the ditch bottom to the road edge were the clovers, timothy, and brome grasses seeded by the municipality. The road itself was a dirt-with-gravel affair, a typical road used by the locals to reach their fields and to haul feed and grain, depending on the season. It wasn't heavily used but was a public road nonetheless and so did see traffic from time to time.

My father and mother, having viewed this unused forage and knowing they had to do something to keep the cows fed, came up with the plan of ditch grazing. It went like this. During the evening the herd was sent out to one of the pastures south of the barn for their overnight grazing. During the day, however, the herd was allowed to go out onto the road and graze the ditches. Can you imagine the exhilaration if you were a cow? For years you had looked over, stuck your head through, and tried your best to get on the other side of the fence. Today your dreams were realized; you couldn't believe your eyes as the farmer opened the gate and chased you into the road ditch. Well, we all know that cows aren't necessarily that bright but they did seem very happy to be out of their paddock. Freedom is enjoyed by everyone, even cows, I suppose.

So the sixty-six-foot municipal right-of-way became my brother's and my place of service on many summer days during those dry years. One of us was placed at each end of the mile and our job was to turn the twenty-five or thirty cows in the opposite direction once they had grazed to our end. This was not a small feat for a small boy, but with a great deal of yelling and arm-waving and running eventually the group was turned and headed the other way. It certainly developed cattle sense and helped you know just how to chase a cow and at what point she would bolt ahead or double back. When running beside a critter there is a certain "sweet spot" where you can keep it from bolting or turning around.

You also learned a lot about the different cows' personalities. They were just like people. Some would get right to grazing and stay put as they filled their mouths and rumens with the succulent forage. Others just had to see what was over the next rise and so grabbed a mouthful of forage and chewed and hiked quickly at the same time. The whole herd, once they were getting full, tended to lift their heads and move along much more quickly, out of boredom or inquisitiveness, I suppose.

The job was a great one. You sat on the side of the road and waited until the cows grazed to you and then turned them and sent them on their way in the opposite direction. Between times you had to stay at your post but there were things to explore, flowers, birds, and insects to observe. My older brother was a reader, and so I'm sure many books were consumed as he awaited the turnaround from his end. Me, not so much. Probably because I wasn't in or had just begun school and so my reading ability was limited. But there were other things. Catching grasshoppers was one of them. One could observe how far they could hop with all appendages in place. Then, with one or two removed, one could ascertain what the difference was. Unfortunately, small boys can be a little cruel doing experiments; however, having said that, there doesn't seem to be a shortage of grasshoppers today. Quite often the cows would scare up nesting birds and then we could observe the nest and eggs and make sure the cows didn't trample them. Flowers were also in bloom. You could take a few dandelions and insert the stem of one into another and make a very long-stemmed flower. Taking these home made my mother exclaim over what a long stem I had found. I'm sure she was never fooled, however.

There were also ant "highways" to follow. Quite often a well-worn path between two ant hills just under the grass undergrowth could be observed. It was common to see birds flying, hawks, falcons, songbirds, waterfowl, doves, and meadowlarks. And if we had a very lucky day maybe a garter snake could be observed and picked up for "further use" later that day. There were also the mounds of pocket gophers, or moles, as we called them. During the night they quite often built a new tunnel underground and pushed the soft earth into mounds above ground. The resulting fresh earth felt so good on a pair of bare feet. Leaping between them became an art. They also made a perfect place to set up a miniature farm using sticks and small rocks, fernlike plants, and so on to develop your yard site.

As previously mentioned, for the most part the road was quiet but I do remember one vehicle stopping and talking to us. They claimed they were doing an article for a farm magazine and that we would soon appear in it. They asked a lot of questions from two small boys and I suppose we answered as honestly and as best we could. It was sort of unsettling in a way because they never visited the yard site or spoke with our parents. We often heard our parents discussing Eastern Communism, and of course we wondered if maybe there was a connection with this questioning we endured from what we thought could be some undercover, covert group.

The ditch feeding continued during these times of drought. It continued in the summers that were too dry to develop adequate pasture and before the herd began to be fed in the dry lot when silage took over as the main feedstock and was fed year-round. Many days two small boys spent turning and returning the dairy herd on the mile in front of the farm. The learnings that the job exposed us to were many. Oh, and by the way, we never did appear in the farm magazine.

# CHAPTER 13

## Introduction to Yellowjackets

The hammer mill sat all by itself between the barn and the house (closer to the barn). The mill consisted of a large metal drum with an inlet and outlet that was mounted on a steel frame supported by a pair of skids. Its purpose was to grind up feed to make it more palatable for the dairy cows. Cows are like children in many ways; you need to "hide the peas with the potatoes," or in their case, the grain with the forage, in order for them not to separate out one ingredient and leave the other. If you didn't do this, cows would waste a great deal of feed over a winter.

And so prior to my father changing to silage as the main forage source, he used the hammer mill to make feed for the dairy cows. It was quite a process. As mentioned, the mill consisted of a metal framework holding a barrel-shaped metal drum that surrounded a smaller drum on a shaft that had metal flails or hammers attached to it. This shaft started at one side of the barrel-like drum with the drive belt shiv and continued through the flail drum in the main mill area and passed into a fan attached to the far side. In other words, one shaft went through the whole device and had all the working parts attached to it. Power to run the operation was supplied by the belt drive on the tractor of the day, probably the 15-30 McCormick Deering.

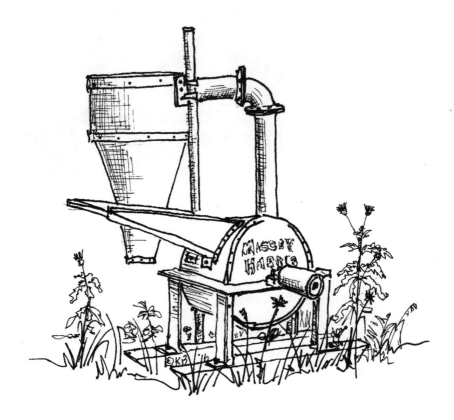

The grain and bundles were fed into the mill through an elongated hopper that was attached to the steel drum in the upper third of one side. It almost looked like the snout of an alligator with its mouth open and the upper jaw missing. In the throat of the hopper were metal strips hanging down from a rod at the top with the purpose of not allowing the mill to regurgitate the feed fed into it. A rounded screen with holes surrounded the hammers and, depending on the size of the holes, gave a coarse or finer grind to the resulting mix. Under this screen portion of the mill the bottom was steeply slanted to allow the ground material to exit into the fan that then blew it through a five- or six-inch pipe up and over to a cyclone separator. This device was used to separate the air and the ground material. The fan blew the mixed forage into the separator that again was like a portion of a barrel with a ten- to twelve-inch hole cut in the lid. The blow pipe entered the barrel on the side and was positioned so the material began to swirl around; as gravity began to pull it down it entered the funnel portion of the separator and continued to swirl until it dropped out the opening in the bottom. The whole purpose of the cyclone separator was to separate the material being moved by the action of the fan from the air doing the moving. As mentioned, the hammered material dropped out the bottom of the separator while the air blasted out the top. This air blast didn't carry any mixed material, but a great deal of dust left and usually covered a sizeable area around the operation. In the winter it was an especially great place for the redpolls and juncos to hang out and pick up any morsels that had blown onto the snow. It wasn't hard to tell if Dad had been grinding feed.

I do remember watching Dad grind feed with this device. As with anything mechanical, there is a technique. I remember him laying an oat bundle on the feeder throat and cutting the string and then feeding portions of it into the mill. The twine may have been ground with the feed or saved for various jobs around the farm. It was made of sisal, a natural product from the Agave Sisalana family of plants (from Mexico) and I'm assuming wouldn't hurt when ingested by a cow. I can imagine, from later things I've learned, that wet material didn't move through the pipes so well and I'm sure it would have been a brute to unplug. At any rate, for a while the mill was used to chop the forage for the cows. Keep in mind that this was all moved by hand, as there were no such things as tractors with front-end loaders in those days. So the bundles would have been brought to the mill on a wagon and the chopped forage piled up under the separator and brought to the cows by wheel barrow. No wonder Dad developed the pit silo with a moveable head gate, so the cows could partially feed themselves.

Even though the hammer mill wasn't used very long for chopping forage and grain for the cows, I'm quite sure it had its day chopping grain for the hogs that had previously been housed on the farm. Hogs require well-chopped grain to help them digest and get the value from the grain. Being mono-gastric, they only have one stomach to get the job done. Cows, on the other hand, require grain to be chopped somewhat but the action of the reticulum, the first of four stomachs, helps with digestion of heavier materials such as grain. They don't require a fine chop like hogs. As I recall, the old mill stood with a few pieces of abandoned pipe leaning against it and the cyclone separator on the ground lying on its side. It was a perfect place for large ragweed and pigweed to grow. Remembering the dust that had settled around the area each time feed was chopped, it was also an area that contained great nutritional value for whatever seed decided to sprout there.

Midway through the summer the weeds had grown very well. On my way by the mill one day I decided to pick a number of these. As I pulled them up I noticed a large chunk of dirt, which I decided to knock off by flailing it over the old mill. Big mistake! You see, a large group of yellow jackets had decided to build their hive in the unused mill. I flailed away and suddenly was overcome with a large contingent of yellow jackets. There were wasps everywhere. They began to sting me all over my face and arms. I didn't lose any time but ran to the house hollering and crying out with every step, still flailing my arms trying to get away from the carnage. My mother heard me coming and met me at the door. "What happened?" was her first urgent question. I managed to get out some garbled response about wasps, being stung, the barn. "Where did you get stung?" she inquired.

"IN THE BARN," was my emphatic response.
"But where on you?"
"IN THE BARN," I responded. After a while she managed to get me calmed enough to hear the story of what happened and the reason for my pain. I learned that day that I wasn't allergic to bee or wasp stings, as no anaphylactic reaction occurred. However, when the count was in I had sustained twenty-three stings all over my face and arms. My mother's nursing instincts took over and I was plastered with baking soda poultices and probably relieved of any chores for the evening.

I do remember the next morning; I could barely see. Both eyes were swollen shut, my face was swollen; I looked like a prize-fighter. I'm sure my father was dispatched promptly to wreak havoc on the occupants of the old mill. In the end, however, I did heal and I gained a new respect for hornets. The story of my cries of "IN THE BARN," and "But where on you? IN THE BARN," were told and retold at family events and gatherings to the enjoyment and hilarity of all who were listening. It took a few years, but as I look back even I have found the response rather amusing. There is, however, still a little angst as I recall the meeting between a small boy and a nest of angry hornets.

# Chapter 14

## Mona

Her name was Mona. At least that was her stable name. As she was a registered pedigreed Holstein-Friesian, she would have had a prefix name and probably a second name that included some of her lineage, something like Spring Ridge Rag Apple Mona. She didn't carry the Poplar Glen prefix as she had been purchased from Bailey Farms at Clover Bar. She had originated in Ontario and had arrived in Alberta with a shipment of heifers. Dad was buying some heifers and had gone to Bailey Farms where some were for sale.

Mom and Dad figured that if you were going to milk cows they might as well be good ones. They were pleasant to look at and came with a pedigree that when traced gave you a bit of an idea of what to expect in their offspring. It then became an art to read the bull catalogues and try to find the best match for the heifer and pedigree that you were working with. My parents, being leaders in the use of artificial insemination, were on the cusp of this match-making. And over the years it paid off with high-producing and good-looking animals. A pedigree like "stood next to an award-winning, high-producing cow on the west side of the barn" just didn't cut it. There was excitement and anticipation as one of your good females delivered you a heifer calf. The warm feeling of "all is right in the world" certainly accompanied one of these events. Of course, there were also the bull calves born or the losses of offspring that tended to dampen total euphoria.

And so Mona arrived at Poplar Glen. She was a beauty. The contrast of the coat opposites of black and white is stunning. She was more black then white, as I recall; however, there were large white pages as well and a partial white blaze on her face. And she was smart. For those that have never worked with cattle one cow appears to be the same as the next. This certainly is not the case; I learned many times over the years that cows are very much like people, with definite personalities. Mona was interesting, to say the least, and was all personality. It didn't take long for Dad to find out why such a good pedigreed heifer was for sale for such a reasonable price. Mona was her own cow. She didn't like to be penned up and thus could open most gates and restrictions that were there to keep her in. No wonder she had been tied in the barn at Bailey's. If there were cow popularity contests, she would have been a winner just because of her ability to let everyone out overnight.

It wasn't long until Mona had taken a shine to the children on the farm. You could go into the cow yard behind the barn and Mona would stop what she was doing and amble over for a visit and a scratch. She particularly loved you to scratch the poll of her head. I'm not sure who she thought we were exactly but it became her job to protect us. If the current farm dog accompanied us into the pen, she made short work of chasing it out again and then would come back to be sure we were OK. I can still picture the dog making for the fence at a full sprint, slithering between the bottom rail and the ground just ahead of Mona's snorting muzzle. She would return and sniff and lick us as if to say, "Are you OK?" Of course, one of the most motherly things a cow can do is lick. The first thing she does after her new calf is born is to lick it all over while making soft, deep-throated, matronly mooing sounds. Her affection was definitely showered on us. I remember letting her lick my hair. The strength of her tongue wrapped around my hair and pulling along its length felt very good, as I recall. I used to let her lick for awhile until she tired and started on the side of my head and face. The line had to be drawn somewhere and so the session would be over at that point. Her tongue was so raspy that if she continued licking it felt like the skin could be ripped off. Naturally one had to have a hair wash in the evening, but for the day you had a nice "cow lick."

Cows are ruminants with a vast second stomach that has a process whereby forage and other such material can be digested. In this vast "drum "of microbial activity this is broken down to free up nutrients for the absorption of the true stomachs, the omasum and abomasum. The process begins as the cow consumes the succulent grasses and legumes that make up a vast part of her diet. Without upper front teeth, the cow uses her tongue to wrap around the forage and rip it off and begin to chew. This first processing doesn't take long but happens as she rolls the forage around in her mouth, forming a sort of ball. This ball or cud is swallowed and ends up in the rumen, where it is preserved for a later time. Then when the cow is full she will find a comfortable place to lie and begin to regurgitate cud one at a time, re-chew until it is finely ground, and then re-swallow. At times I would lie on the ground and look up into Mona's mouth as she lay chewing her cud. While chewing her lips would part slightly with each sideways chomp and if you looked up in her cavernous mouth you could see part of the cud as it was being ground more finely with each crushing grind of her mandibles. This would carry on for a minute or two and there would be a pause as she swallowed the ground-up cud. If you watched her throat you could see a lump travel down her esophagus and disappear at the base of her neck. In a short time you would be able to see the reverse happen as a lump travelled up her neck and the whole process repeated itself.

The action of the bacteria in the rumen and the re-chewing of the cud renders the contents suitable for absorption as they pass into the true stomachs and intestines. The value of ruminants such as Mona cannot be underestimated; they are extremely important in turning feedstuffs that are not normally useful to mankind into food easily digested by us, such as milk, cheese, and meat products. The world would be much hungrier without all the Monas out there.

So Mona would continue her cud-chewing. She would lie comfortably with her eyes half closed, a constant chewing motion, the picture of contentment. I would sometimes find her in the group and lie with my back against her massive side, place my hands behind my head, and just sit there feeling the motions of her chewing and breathing. It was a comfortable place to be and to dream in the sunlight encircled by the warmth of her immense body. There was little doubt that she enjoyed it too.

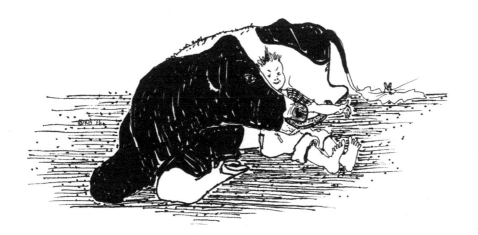

# CHAPTER 15
## Driving Cows to Pasture

Pasturing the cows sometimes included driving them to Grandfather Dixon's place. A corner gate was made on his and Grandma's half-section of farmland and my brother's and my job was to herd the cows along the road in front of the farm to this pasture. This was a very interesting job indeed, especially watching the different cow personality types show through. There were the girls who had to be in the lead, always looking for what lay ahead. There were the ones whose interest could easily be piqued or altered by the flutter of a bird out of the grass. Others were content to plod along as observers, it seemed, and still others were followers, just doing what everyone else did. In some ways it did remind you of the "Ladies Aid Society" with its variety of personalities and dynamics. Perhaps that is why we usually referred to our herd as girls or ladies rather than cows.

Mona didn't seem to fit into any of the ordinary cow personalities. For the most part she thought she was one of us and not one of them. Her method of travel to the Dixon pasture was to begin grazing as soon as she arrived in the ditch after departing the barn enclosure. She paid great attention to this task. After all, she had just been milked and was very hungry and the succulent grasses and vegetation were to be enjoyed as she slowly meandered along. This she continued to do until the rest of the herd was out of sight and then she would raise her head and with great concern begin to race until she caught sight of the rest of the bunch, whereupon the slow grazing would repeat itself.

This wasn't a big deal except for one thing. As mentioned, Mona enjoyed every moment of the attention we gave her. And as the pasture was a mile from home, which seemed like a long way for two little boys, we thought maybe a ride would be in order. Mona was okay with it and so we would ride her as the herd wound its way to the Dixon place. But staying on her as she played catch-up to the rest of the herd also made for some great rodeo rides. I'm sure we didn't always stay on for the required eight seconds of rodeo conventions but it was just a matter of getting her close to the ditch-cut bank and interested in some forage and then we were in a position to remount. I can still see her turning her head to her side, smelling and licking our bare feet as we sat on her back. All was well and good until the next catch-up.

And so the days passed as we drove the cows to the Dixon pasture. It was easy to tell the time of year we were in by the vegetation, flowers, and bird life encountered as we moved the herd along. The early spring arrival of the prairie crocus was usually not part of the pasture drive as our own home pasture was usually adequate at that time, providing we had a good snow cover or spring rains. But sometimes in June when the buffalo beans were in full bloom, the dandelions and silver willow were showing their finest, and the whole prairie scene was experiencing its time of renewal, we would be on the job of moving the girls. It was always interesting to make note of where the saskatoons, pin-cherries, and chokecherries were blooming along the way, as there would be a chance for us to feast later on in the summer and fall. Added to that were the wild strawberries and raspberries that showed up in the lighter soil exposed by the road-building.

During the hot lazy hushed days of summer, there were times when it seemed not a breath of air was moving and the sun's heat was being trapped in the road ditch. It was such a relief to have the cows following their path along the ditch bottom because of the recovery of the small aspens that had once again taken up residence in the ditch. They were big enough to provide a nice cool shade and relief from the direct sun. These saplings were covered with leaves and as you passed under their umbrella, the coolness was most refreshing. Their branches also moved along the face and swept away the horn flies that were a constant irritant during those hot summer months. And bother the cows they did. Mona would throw her head towards her shoulders, getting some brief relief from the constant fly irritation but in the process covering the front of her back with saliva. Then she would repeat the process on her other side. Naturally the flies loved this as there was more material for them to devour, displayed on her back in the warm sunshine. None of these pests, it seemed, were anxious to hang around Mona's tail end, for one swipe of her tail could be a very destructive experience for a fly.

Sometimes we would see cowbirds and at times they would land on Mona and her compatriots, feasting on the flies. Often these birds could be seen as they accompanied the grazing herd during the day, mostly on the ground between the cows but sometimes on their backs as well. Both sexes were involved in this insect hunt even during their family-raising phase, as they had developed the practice of "nest parasitism." These birds evolved following the great bison herds for eons and because of their nomadic nature would lay their eggs in the nests of songbirds and let them raise their young. I remember the time that my father found a nest that had been parasitized. The cowbird had laid an egg in the nest and the songbird had recognized this and had rebuilt her nest over top of the original, leaving the first-laid eggs and that of the cowbird below in the lower level.

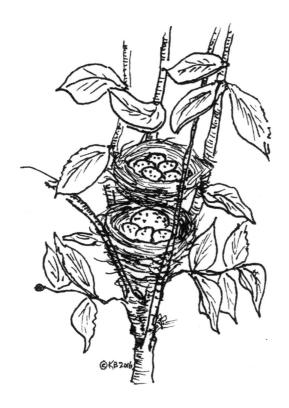

Many insects were encountered as well: the grasshoppers visible on the pocket gopher mounds as they laid their eggs in late summer, the sound of crickets as they scraped their hind legs together to make a ratchet-type racket, the wasps and hornets as they developed their grey paper-like nests that grew as the hives grew, and of course the ants. Two medium-size types seemed to be the norm: the red and the black. It was fun to follow their trails underneath the prairie wool from anthill to anthill. It was amazing to us the amount that one little ant could carry.

Stockmen are ingenious in trying to train animals to remain where they are supposed to. One invention, the "cow poke," was a device that fit over the neck of the offending cow. Its purpose was to catch on the upper and lower wires of a fence as the cow wearing it was reaching through and trying to get feed on the other side of the fence. The pressure of a cow reaching through a barbed wire fence eventually loosens the staples and the wires then drop, leaving free access to the forbidden area that the fence was protecting in the first place. I can't honestly say I remember one of these on Mona, but there could very well have been, as she was always looking for the next way out of her enclosure.

She was clever. We milked in a three-in-line milking parlour, simply meaning there were three milking stalls in a row where the cows stood head to tail. These were individual stalls with each one working independently of the others. The stalls were raised so that the waist of the operator of the milking parlour was even with the floor where the cows stood. At both ends of the three stalls were a set of operator steps to go up or down to do whatever needed doing while handling the cows. Behind the three stalls was an alley where the cows could come into and exit the stalls. An entrance door and an exit door were at the end of this alley. The doors were held up by barn track and rollers and sloped towards the closed position. A rope on the door passed through a series of pulleys so that you could open the door from the milking parlour pit. The door opened by pulling on the rope; when the rope was released the door travelled back to a closed position because of its weight, with gravity sending it down the slopping barn track to rest in a closed position. Here is where the ingenious Mona was at her best. Mona had observed that the doors could be opened by using her nose to push the door up the rail and barging her way through; she thus could enter the milking parlour at her leisure. She must have become bored in the evening or at night, for it was at these times she checked out the entrance door to the parlour and thus entered. How she developed the next part of her habit I will never know. The power of observation, I guess. She would sneak down the operator stairs and open the sliding door of the chop bin that was located along the pit wall opposite to the milking stalls. Although getting into the bin was a clever act on her part, loading up on chopped grain wasn't. It can cause an illness know as "grain overload" in cattle, whereby the bacteria of the rumen are killed due to the acid buildup of the consumed grain as it sits as a rotting mass in this organ. If it isn't removed in time the toxins can kill the animal. For some time Mona didn't overload on the grain. On a number of occasions, coming into the milk room attached to the parlour, you would hear a scrambling of hoofs, a sliding and banging of either the entrance or exit door, and all that was left as evidence that someone had been there were the hoof marks, indentation in the chop bin where some had disappeared, and often a cow patty where Mona had made room for the evening feast.

I don't remember how Mona's demise came about. I don't recall for sure whether she had any female offspring; I don't think so. It could very well be that she overloaded on grain one too many times. I do know that the driving of the cows to pasture not only accomplished the feeding of the cows during the summer but gave two young boys the opportunity to learn a great deal about nature, about cows, and the smells and sounds associated with both of them and their surroundings. Mona will always be with me even after these many years. She still travels the pastures of my mind.

# CHAPTER 16
## Oh the Aromas

The power of aromas to trigger the memory certainly has been my experience. Let me lie down on a nice piece of grass with the sun warming my body, my eyes closed, a slight breeze rustling the slender waving fronds of goldenrod or gently moving the stocks of the various grasses that make up the grassy mosaic, and I'm back to my childhood and the many fragrances that were part of it. It's amazing the flood of memories that accompany them.

I love the smell of fresh-cut oats. It takes me back to a time when, as a small boy, I was riding along with my father in the late afternoon as he cut the first round of an oat field. We were near trees along the south side of the west quarter section and Dad was making bundles with the binder being pulled by the little Ford tractor. This was quite an advancement because the 8N Ford had a power takeoff that was connected to the binder. The table and bundle carrier controls had been moved from the rear of the machine and mounted on the tongue of the hitch. They had previously been mounted at the rear of the machine on earlier binder models. This change made them available to the operator from the seat of the tractor. Naturally, when horses were pulling the machine, they could be driven from the seat of the older-style binder that was mounted on the rearmost part of the implement, next to the bundle-tying knotter and the twine box. This meant the power to run the sickle, reel, and canvases was provided by the tractor and not by a bull wheel and drive chain, which had for years been the source of power for these machines. The bull wheel provided power through a chain engaged on a sprocket on the big wheel that rotated when the binder was moving forward. Naturally, the machine had to be moving in order to rotate the wheel and thus the mechanism. This worked fine under ideal conditions but was a real pain in harsh conditions when the wheel slipped or the machine plugged due to heavy or moist crops. With the newer power binder, it was possible to run the binder when standing still, which was a huge advantage over the older binder models, especially when you were plugged.

As I rode along with my father that afternoon it was a typical fall day. The sky was blue with a few cumulus clouds drifting by. As the afternoon wore along these cumulus began to turn into cumulonimbus, producing the odd shower here and there. It so happened that one of these developed over us and it began to rain and drop a few ice pellets as well. Obviously this was not a new happening for Dad, for he soon had the two of us stretched out in the stubble under the table of the binder with the rain beating on the canvass above.

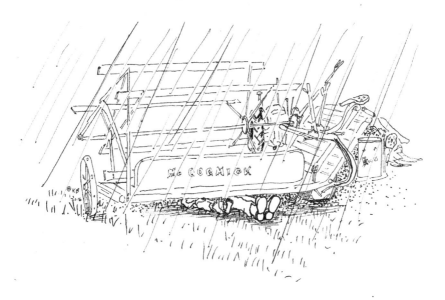

Keith Erwin Brower

The memories of the odours detected and the flora and fauna seen and experienced is still with me. There was the whiff of warm grease that had been pumped into the bearings on the canvas drives and gears prior to our starting. Damp canvas also had a distinct odour, as did the mint-like smell of the weeds that had escaped cultivation and the knife of the binder.

As my eyes glanced along the adjacent fence line, it was fun to pick out and ask about the various bushes and vegetation growing there. Dad was always interested in the history and various characteristics of the flora of the area. Fortunately he was always willing to share and explain the knowledge he had about each. As I gazed along the fence line, it was not hard to pick out buckbrush, with stark white berries, dogwood, with its distinctive red bark and white berries, saskatoons, whose berries had dried into prune-like dollops, and of coarse the beautiful clusters of juicy black chokecherries just begging to be picked and devoured. Dad explained how important these dried berries would be for the wildlife during the winter ahead. Certainly the berries of the buckbrush would be consumed by the sharp-tailed and ruff grouse during these months. Many times since, I have observed these birds walk on the snow among these bushes, with necks stretched as they reached to pick the berries resting amongst the dried leaves on the tips of the shrubs. It wasn't hard to imagine how the opportunist of the plains, the coyote, would stand on its hind legs and strip the limbs of the chokecherries, devouring the berries for a nice tasty treat. It has become apparent over the years that this is quite often the case in the fall when the scat of this canine is observed.

I think the most stark observation from under the binder that day was the beauty of the kaleidoscope of colour that lay just beyond us, both in the fence line and the bluffs of mixed trees and shrubs that formed a palette before us. It was amazing the contrasts that could be included in a single group of trees. In a bluff of poplars, for example, there could be a captivating cadmium yellow clump framed by jade green or brilliant orange. Certainly with this array of colours, enhanced by the mottled browns of the saskatoons, the shades of green found in the prairie grasses, and overshadowed by the wide expanse of the blue prairie skies, my affection for the plains was born. This has swelled to include a love of the fall season. And with the fall comes all the bouquets of scents that conjure up these memories. That's where the scent of freshly cut oats comes in. In my mind it is a clean, aromatic scent, very distinct. It was strengthened that day by the rain shower. It's the same scent that can be encountered as livestock, be they cattle or horses, chew away at their oat fodder. Naturally that also conjures up a flood of memories of different times and different situations.

Another vivid memory deeply embedded in my mind is the cutting and associated aromas from freshly mown hay. It was such an interesting thing for a small boy to watch and enjoy. And oh, the fragrance. The process of mowing hay was not a long-lasting operation on our farm because of the early adoption of ensiling crops. But I do remember watching from the tractor fender as the mower blade and sickle did their work. These could not be seen as they entered the crop. All there was was a sort of tremor as the action of the sickle worked on the standing hay crop followed by the falling over of these same plants. Because of the forward movement of the mower, the plants all fell over butt first with the heads to the rear.

There was a very sweet smell as this occurred, produced by the flowers of both the legumes and grasses as they fell. This continued as the hay dried and was raked and baled to be kept for winter feeding. There is no aroma like opening up a fresh bale of succulent hay and feeding it to your livestock. Unfortunately, it was not always possible to get hay up in this condition, especially when it rained shortly after mowing. This could be very frustrating and I'm sure was a major reason for changing to silage in those early years.

Another interesting aspect of mowing hay was that it usually occurred in late June. This was the time when ducks built their nests; quite often they would lay their clutch of eggs in a small indentation in the hay field surrounded by the protection of the standing crop. The duck would pluck feathers from her breast and line the nest with these feathers and other soft materials gathered in the vicinity. Then she would sit on her nest, protected by the camouflage of the surrounding hay crop. Naturally as the hay crop was cut these nests would be exposed. As I rode along I kept watch for a duck that would all of a sudden rush out and fly up only to land a short distance away. Dad would stop the mower and we would look for the nest and clutch of eggs that would inevitably be in the area just cut. We would surround the nest with some freshly cut hay and hope that the duck would return. Quite often she would and live to hatch her clutch even if we had to move the nest during subsequent haying operations. The only problem she had was the constant lookout for the scavengers that were always hanging around, be it the crows, coyotes, or gulls. We weren't always successful in hiding the nest well enough to foil this opportune group.

When I think about it, all the scents of my youth convey special memories, but one was especially unique. The most embedded memory is the scent associated with Uncle Fred's pipe. Neither my parents nor grandparents smoked, so it was a novelty when Uncle Fred would show up, usually selling tickets to the Vermilion Fair, and the aroma of his Sweet Virginia wafted through the house. It usually took Uncle Fred quite a while and a fair pile of matches to get her going but once he did the smell of the burning tobacco seemed glorious. The floating bluish smoke, the smell of the burning leaves, and the ritual of lighting were fascinating for a small child, though smoking never did catch on with me. Maybe it was the time factor required to get the pipe going, as demonstrated by Uncle Fred on many occasions, that contributed to my not taking it up. But oh the aroma; I still love the smell of smouldering, glowing Sweet Virginia good old pipe tobacco.

# Chapter 17

## Arden

The 1939 Chevy, green in colour, a fading green, would slowly chug up the road and come to a stop at our farm gate. Covering the truck bed was a caboose-like structure, also in a fading green, with a stove pipe protruding from its roof and a door tacked on the back covering its entrance. In later years further inspection of this early truck camper revealed a metal five-gallon pail with a hole in the cover to accommodate the bottom end of the stovepipe, a hole in the side to feed fuel through, and an antifreeze pail full of lump coal, old newspapers, and small kindling. This was the camper heater. Across the front was a very old spring and mattress set, rather grungy, covered in old blankets and various other pieces of clothing, assorted extra coats, socks, and so on. An odd assortment of cans, bottles, balls of string, a few tools, and cardboard boxes, amongst other things, made up the rest of the interior decor.

It was spring, and as sure as the ducks and geese returned from the south, so too did Arden Hamilton. He had not actually gone south but was more like a squirrel or gopher; during the winter months Arden did not drive but come spring his truck was back on the road. During the winter he either holed up in his shack or, especially in his younger years, lived with one of many neighbours in the region. Arden was a character, a one-of-a-kind sort of individual. My mother and grandmother told of his family, which lived in the district. Arden was the first of three children, and being a big boy may have experienced a difficult time entering this world. In the years when he was born, there were no cesarean sections; looking back now, it's possible that his mental capacity was affected during his birth. At any rate Arden was not, one could say, normal. He had a few idiosyncrasies that he handled as best he could with the hand he had been dealt. Certainly his siblings did not share his oddities, as I believe his brother became a medical doctor and his sister a nurse. And he had had a good upbringing. He was loyal and as honest as the day was long.

The stories of his younger years as a strong, robust young man are somewhat legendary, as remembered by the matrons of the family. As a young fellow he had been engaged by a certain town citizen to look after his cattle, which were being grazed in a community pasture situated behind where the town of Wainwright's post office is now located. The only problem was that another citizen had engaged a rough-and-tumble young railroader, Alf Lilly, to do the same with his cattle. Each young man's cattle were chased out by the other and replaced by their own charges several times during the afternoon and evening until the two young upstarts came to blows over who should have their cattle in the pasture. I understand it was quite a skirmish, starting at one end of the back alley and continuing for several blocks with no declared winner. That may be because the incident was cut short by the local constable, who took the two combatants to the local "clink" where they spent the rest of the night cooling their heels before being let out in the morning. I don't recall ever hearing whose livestock was allowed to graze the community pasture in the end.

During those years Arden worked off and on for different farmers in the area, including my grandparents, who sometimes during the winter months would head to White Rock, B.C., for a chance for Grandpa to get back to saltwater for a season (he had been raised on the Richibucto River in New Brunswick). Arden would be engaged to look after the livestock and house while my grandparents were away. He would show up with all his necessities, which included a vast array of self-medication he religiously took. The windowsill would be lined with bottles and tins of varies colours and sizes. Such monikers as Sloan's Liniment, with the good Doctor Sloan himself prominently displayed and looking dapper on its label, red hot Mustard Compound, Carbolic Salve in its yellow tin, and Absorbine Sr. with the picture of a muscular heavy horse adorning its front, all wrestled for a position on the ledge. Not to be outdone were the bottles of Dodd's Kidney Pills, which made your urine blue, and the bottle of Sweet Spirits of Niter, another product often used for the kidney ailments of horses.

It seems that Arden had at one time or other had been warned about catching cold, perhaps by his mother, and so he always wore several layers of clothing, summer or winter. He also believed everything he heard about ailments and soreness and was always treating some ailment or other with quack medicines, self-prescribed. Thus the array on the windowsill. My mother's skills as a nurse translated into long phone calls and discussions in later life as Arden unloaded all of his various ailments and medical problems on my poor mother.

When engaged to look after my grandparents' farm, Arden would move in and have a cot by the cookstove in the large farm kitchen. My grandmother required that he go to the woodshed and hand out his clothes to be washed while he took a bath in a provided washtub. This was a ritual followed by the women of the district whenever Arden was a guest workman at any of their farms. The clothes would be returned after hanging to dry and the next set would receive the same treatment. There was no shortage, as he peeled the layers of clothing off. Usually after one of these visits he left cleaner than he had been for months.

It seems that the families of the district would give him a bit of cash out of their meagre earnings but for the most part he was content with a place to stay, some gas for his old truck, and canned saskatoons. According to my grandmother, if she had canned saskatoons, there wasn't anything that Arden wouldn't attempt to do. He loved them.

Certainly as a younger man Arden was able to find work, for he was an excellent fencer. One of the mechanical geniuses in the area, a man named Guy Gibson, had built a saw platform on a wagon for Arden. It consisted of an old Rumley single-cylinder engine mounted on the front with a large circular saw mounted with pillow blocks on the rear of the outfit. A flat six-inch drive belt ran from the engine drive pulley to the pulley on the saw shaft. I remember seeing Arden start the rig. It seems that he used a battery to fire the magneto while he cranked the rather large engine. Once he had it firing he needed to transfer the spark plug wire to the plug itself and I distinctly remember how it jolted him until he had the wire attached to the spark plug rather than being grounded by him. This was all endured in order to get the engine running. Then there was a constant "bus" and "putt, miss, putt, putt, miss," as the engine fired and ran the saw.

Once he had this rig, Arden and his friend Billy would spend the next month to six weeks making fence posts. He was able to go to different farms and find a grove of the right-sized trees and the work began. As you can imagine, this was not easy work either. This was before the chainsaw. The trees were cut by hand and hauled to the sawing outfit, where they were cut to seven-foot lengths. Next they were sharpened at one end by shoving an end into the saw at an angle two or three times, rather like sharpening a pencil. Once this was done the post was peeled up a couple feet with a one-and-a-half-inch strip all the way up. After these had been stood up in a tank of bluestone water for a few days the posts were ready to drive. So for quite a few years Arden put in a lot of miles of fence in the district: good fence.

As time passes for us all, we are not always able to do the things that we did as younger people; it was no different for Arden. When I was a boy, he was past his prime. My memories are those of him as a person who was feeling some of the problems of age.

So the door of the old Chevy would slowly open and a hefty individual would slide out and his feet reach for the road. The old seat would have some sort of old blanket on its springs, these being visible from the side where the cover was completely worn off. The pockets in his large long-cut mackinaw bulged, his black wide-brimmed hat was pulled down to meet his ears, and he started with a brisk walk down our lane. The closer he got to the farmhouse, the more he slowed and began to limp to show what bad shape he was in. With Mom's nursing background, there were many discussions about his ailments. His explanation of what was ailing him quite often included the term "buckled nerves." He would hook his two pointer fingers (soiled pointer fingers, mind you) together to demonstrate just what was happening in some part of his body. It seems his information and diagnosis came from an interesting source, a certain lady named Grace Thane of Midnight Lake. I've often wondered over the years just who she was and what sort of an organization she ran. According to Arden she hung a string with a weight on the end from the ceiling or joists above and had him back up to it to demonstrate that his spine wasn't perfectly straight. Doesn't sound like the Mayo Clinic, but enough I'm sure to separate Arden from some of his disability pension.

The only way to get him off his aches and pains was to talk to him about the time that he had travelled to the coast. Why he ended up there I don't know. Some kind tugboat captain had let him see the engine and the engine room on his ship. Arden could recite everything about the engine displacement, thrust, RPM for maximum horsepower, and fuel used. He was totally engrossed in how the tug worked. Another source of interest was the well-worn article he had that pictured and talked about the transformation of a WW II tank into an agriculture tractor. He would talk about this conversion at length and unfold the article with its well-worn creases, threadbare from handling.

At any rate, Arden had arrived with pockets bulging with his spring deliveries. He had taken up with the Cressy Spice and Product company and for a number of years he would place an order to be delivered every spring. It must have been interesting at Cressy's headquarters when his order came in. The only two products that Arden ordered and sold were black pepper and pure dark vanilla. Several boxes would be delivered and Arden would tour the country, visiting and selling these two products until they was all gone. The good ladies of the district would faithfully buy a bottle or two of vanilla as well as a couple of cans of the pepper, just to be of help. I'm sure, in most of their cupboards, there were ample supplies of both; I know when my mother was done cooking there was enough pepper and vanilla to start a small warehouse. My grandmother was also a faithful customer and being an old schoolteacher had fun with Arden as he added up the bill for her purchases. Although his schooling was limited and his reading halting, Arden could add up a triple column of figures quicker than a calculator. It was amazing to watch. No sooner had the figures been written down than he had the sum calculated. Grandma would tease him by saying he couldn't be right and she would painstakingly add the columns only to come up with the same answer. How he did it I have no idea; I'm not sure if he knew, but it was captivating to say the least.

One of my final memories was the day my father, brother, and I visited Arden at his shack. I'm unclear as to why but I remember being there and several observations left a lasting impression. It wasn't hard to see why his back bothered him. His bed, at the far end of the hovel, had a mattress and set of springs that were "shot," to say the least. The middle sank to within a couple inches of the floor. Obviously the resident rodents appreciated the arrangement, for their work on the low-hanging mattress was very apparent. His table, where he clearly ate his meals, was piled high with all kinds of different household paraphernalia. He burst out with his cackling laugh when he pointed out that the reason his plate was turned over was to keep the mice from licking it clean. He also noted that he lost a few in his water pail fairly often. When he offered to put some potatoes in the used antifreeze tin can on the stove and put a little dinner on for us, I appreciated that Dad spoke up and thanked him for his generosity but told him that "Jeanie", my mother, was waiting for us for dinner and we mustn't disappoint her. Arden understood completely because he definitely didn't want to disappoint her either; after all, hers was one of his favourite places to stop and visit and get a wonderful home-cooked meal.

# CHAPTER 18
## The Fencing Project

Fate has a way of bringing the best out in people. In Arden Hamilton's case it was an accident that my father suffered that enabled him to change his focus from his aches and pains to those of my father. His visits continued all spring, summer, and fall; I suppose they fell about once a month. After all, he had a large number of people he dropped in on; the farm women of the district, for the most part, were very tolerant and kind. In most cases if he timed things right he could be in one of these farm homes for dinner or perhaps an evening meal.

I distinctly remember how my mother handled the situation. As our family grew and Dad had hired men, there usually wasn't much room at the table. After all, with all the heavy pants, his double red-and-black checked shirts with his "long johns" sticking out past the sleeves (rather grey-looking "long johns," I might add), his two pairs of bib overalls and felt boot inserts, Arden took up a fair amount of room. My mother would dig out the card table and put a chair at the end for our guest. Arden was a hardy eater and could put away quite a plateful. I still recall him wanting a quart sealer full of water with a fresh raw egg cracked into it for his accompanying drink. He said it was a great tonic for him, but somehow I didn't relish the idea and it never caught on with me nor the rest of the family.

Certainly our entertainment for that meal was listening to him talk. He could even get words out with his mouth full—well, sort of—another good reason to sit him at the card table. At any rate, he would talk of all the gossip he had gathered at similar settings across the countryside. It was sort of non-stop, as I recall, but quite entertaining at times. We gathered that there was a bit of "leg pulling" with Arden because some of the stuff he repeated couldn't be true even though in his state of mind he believed everything he was told. It made me wonder what he might say to the next group about us.

He was a character, sitting there sweating profusely, his hair cropped brush-cut style, his horn-rimmed glasses mounted near the end of his nose and with a few whiskers here and there that had survived the onslaught of the razor to live another day. His jowls were rosy red and if he had had long hair and was cleaned up a bit, he was sort of Santa Claus-like.

Unfortunately, sometime during Arden's life, through whatever means, he was admitted to Alberta Mental Hospital at Oliver (just north of Edmonton). Whether the folks that had him committed wanted him there to see if he could be helped or just didn't enjoy having him around, I don't know. I do know that a neighbour from the Gilt Edge area checked him out and he was reinstated to his proper place, doing what he had always done. It certainly gave him more topics to talk about. The one that I particularly remember him prattling on about was being fed "mouse-tail soup." I'm assuming that it was probably a soup that included spaghetti as a base, but not to Arden; it contained mouse tails. Knowing the familiarity he had with some of the residents of his shacks, he could have been an expert in these matters, but I'm still sticking with the spaghetti theory.

So it was on one of these stops that Arden learned about my father's extreme difficulties. We had been chasing cattle in the corral. One of our early front-end loaders was in the corral with its bucket raised and a logging chain dangling from it (that had probably been used to move a cattle-working chute). My father was running hard in front of the tractor trying to cut off an escaping steer when his inner leg caught on the grab hook of the dangling chain. It sliced his leg and hooked onto the sinew and flesh of his inner thigh. In a split second he was upside down, dangling like a side of beef at the butcher shop. He had to pull himself up with one hand and unhook himself from the chain with the other. As you can imagine, this was a serious injury. He was very fortunate that he didn't sever a major artery and bleed out right then and there.

Added to this complication was the fact that Dad was a singer and one of the neighbour ladies had convinced him to join the group that was busy putting on the musical "Pirates of Pensanz" in our community. Naturally Dad had a rather important singing part and I do recall being at that evening's opening performance. Dad was well-drugged and walking with a cane, limping through his singing parts. As with this kind of injury, the swelling and pain were worse and not better for the next few days. It became very obvious that the half-mile of fence that he was about to undertake on the west quarter wasn't going to get done anytime soon. This, of course, was a blow because the land was needed to pasture the cow herd and without this fence this wasn't possible. After all, forage for the cows was always at a premium, it seemed.

And so on his visit Arden heard and comprehended my father's injury story and the situation of the cows that needed pasture; that's all that was needed to put him into action. I recall him sitting on the edge of his chair and lamenting the whole situation with quick clipped words while he rocked back and forth, thinking about what he could do to help, uttering sounds mixed with sympathetic words almost as if this was his own accident. And then it hit him. He knew fencing, and he would complete that half-mile of fence for his good friends Jean and Barry. And so into action he went. His limp and slowness were gone as he hustled up to the old Chevy; off he went to retrieve his tools to begin his work. He was not a young man at this point, well into his sixties I would think, but if there was one thing that Arden knew, it was fencing.

It wasn't long until the old green camper was a fixture in our yard and the work began. I do recall being out at the site a couple times, probably with my mother bringing a drink and hot fresh buns with cheese or some other tasty delight for a break. I watched as he stepped off a good strong five steps, punched a hole using a crowbar, and finally stood on a five-gallon pail and with the ten-pound maul drove the sharpened post into the ground with several well-placed blows. He kept his line of posts straight because he had stretched a single wire the length of the fence between both corner posts and as he punched and drove he used this as his guide. And did he work hard; he moved along and drove posts like he was a machine on a mission. He must have ached at night but I don't recall one bit of complaining. My mother probably picked him up for or delivered a nutritious noon meal while his morning and evening nourishment were taken in our farmhouse. He gave a report of his progress to my father each day—not that my dad was worried, because he knew Arden was a fence expert. He understood that the wire should be stapled to the inside of the posts, animal side, so when cows put their heads through the fence for the "greener grass" (as they inevitably would) they were pushing against posts and not just staples. He also knew about straightening and reusing staples as well as splicing barbed wire. A pail was kept by all farmers as they removed old fence lines and the staples were carefully straightened, as they were used in the new fence so as not to have to buy too many new ones. This was not being cheap, but frugal. Every penny counted.

So the routine repeated itself over a number of days. A good strong fence began to rise on the north side of the quarter and it wasn't long until Dad was driven out to give it his inspection and blessing. Arden was justifiably proud as he and Dad surveyed the straight, strong, and well-stretched fence that stood before them that day. It would take a super-cow to get through it in the short term and as I recall that fence stood for many years as a testament to its builder.

As I think back on this incident, I can see that life lessons were demonstrated to me by the neighbourhood character, Arden. You see, my father was unable to pay him; he refused to take anything except the meals and friendship he received during those few days. However, my family and I received payment in lessons learned during that time. Family, faith and prayers are important, friends are important, kindnesses rendered sometimes return in peculiar ways. I learned that when we fix our minds on others and their needs, our needs seem to be minimized and sometimes completely disappear. My thanks to a simple, challenged man, Arden, who demonstrated this to me so succinctly so many years ago. These lessons are still applicable today and have stuck with me ever since they were taught through the building of one half-mile of fence.

# Chapter 19
## Winter's Fuel

Cold winter weather was one thing that we accepted as part of being raised on the prairies. Keeping households and livestock warm and watered through these winter months was a challenge but nevertheless an important function of survival. Fortunately, coal beds were laid down in Alberta during bygone eons, and they were being harnessed by companies to supply energy to anyone that needed it. Organic materials plus pressure and heat developed the coal beds through a process called coalification. It was some of this bituminous and sub-bituminous coal that we acquired and used on our farm.

Prior to my parents having a truck of any size of their own, a local trucker, Marcel Touchette, was hired to haul the coal from the mine at Forestburg to our farm. Two types were delivered: large lump coal for the tank heater and my grandparents' house, and stoker coal used in the Booker furnace in our home. The lump coal was large lumps of the shiny black substance usually no more than twelve inches across and the stoker had been crushed and screened to nuggets no larger than a couple inches. The stoker was usually sprayed with a light oil to keep down the dust, which could certainly be a problem if the oil was not applied. I remember the coal arriving and Mr. Touchette unloading the truck. He would back the truck up to the window that entered the basement of the house adjacent to the coal bin. This was a cordoned-off area with walls made of planks or slabs and a doorway that had boards added to it in a horizontal manner. These were removed from top to bottom as the coal bin was emptied.

So the task of unloading the coal began. He would start unloading whichever coal was closest to the endgate of his grain box. This was all done by hand with the hoist of the truck raised slightly to give a bit of a slope to the box bed. Marcel used steel shovels to unload; heavy steel shovels. I remember at times he brought his aged father along, who had no doubt gone along to the mine for the ride. He also helped unload and it was most curious to me because with each shovelful he dumped down the coal chute, he let out a gasp-like grunt as if it was the hardest shovelful he had ever handled. This process continued until the truck was empty, every shovelful of a full grain box handled by hand. No wonder Marcel looked like a prizefighter; I don't think there was an ounce of fat anywhere on his body. As I recall, he was a very polite and gentle man who appreciated the business we had given him.

Fairly early on Dad was able to acquire his own three-ton truck for all the duties that a growing farm required of a heavier truck. It was a Dodge army surplus gravel truck. Having the military base at Wainwright had given him the opportunity to apply through the Crown assets disposal division and bid on a truck. I don't know how much he paid for it, but I do know that these trucks usually didn't have a lot of miles on them when they were disposed of. Also, the army didn't like to overwork their equipment. This truck had a very small steel gravel box when it arrived. It probably had one-half the carrying capacity in that box that the truck was designed for. At any rate, the truck was acquired and the retrofit to a farm truck began. Dad removed the small steel gravel box and I believe took the truck to a neighbour to have it lengthened in order to accommodate a newly crafted longer grain box. The khaki colour of the military was replaced with a deep blue on the cab as well as on the box and Dad had a very serviceable truck on his hands.

With his own truck, one of the important jobs that could be handled was the run to the coal mine to procure the coal to keep the different stoves and furnaces providing heat during the winter months. Lumps of coal were important to "bank" or hold the fires overnight. This required adding a lump of coal to the fire prior to retirement and letting it slowly burn over the night, leaving a nice heap of hot embers for the morning kindling and sticks of wood to be introduced. This was great for a quick flare-up and the heating and cooking cycles began all over again the next day.

So, with the "new to us" truck and the need for winter's fuel at hand, the day was set to make a run to Forestburg for coal. The mine at Forestburg was a surface strip mine with the sub-bituminous coal not too far below the surface. It was an important suppler of coal for the local towns and farms, coal being the energy source for heating and cooking in these locations. The use of electrical energy also ramped up in both the urban and rural areas in the early '50s, so the mine became very busy with new contracts to supply coal for power generation.

As our farm was in need of a winter's supply of coal, we headed out to Forestburg. I'm supposing I would have been around four years old and was lucky and willing to accompany my father to the mine. I'm sure it was a long trip for a little guy but, as I recall, I didn't complain as I was delighted to be asked to go along. Dad was also known to find enough change on such a trip to buy a chocolate bar, so one could only hope. It was also an opportunity for me to have my very own lunch in a sugar bag that my mother had salvaged from her pantry. These were cotton bags with "Rogers Sugar" stamped on the front; they probably held ten pounds of sugar when they were new. They were very handy for various uses around her kitchen, including lunch bags. With this in tow for Dad and me, I felt very mature, just like my big brother who left for school each day with a lunch bag like this to go along with his books.

The roads were all gravelled at this time. Pavement hadn't spread much beyond the major centres, which, when you think of it, weren't all that major at that time either. So it was probably slow going and certainly an all-day trip. I do remember being at the mine, with its buildings, coal conveyors, and chutes. There was a scale to weigh the empty truck to get its tare weight and then a place to back under a chute to have the coal run into the truck box. With a changing of position to another loading dock, the other size of coal would be loaded. The truck was then re-weighed and the net weight established to be used for the purchase of the coal at so much per ton.

It was fascinating for me to see the large equipment that brought the coal for sorting and crushing from the draglines out in the field. The other fascinating thing was how everything was dark grey or black. A fine dark dust was over everything. The coal was mostly dry coming out of the ground and as it was transported, sized, crushed, and so on, it became a bit pulverized and airborne. The buildings, the covered conveyors, even the old posts that held the yard lights were all very dark with the remnants of years of bombardment with coal dust. The miners' and yard staffs' clothing also looked rather dark and soiled from the coal. Even the roadways were black with a layer of small nuggets and crushed coal covering them from side to side.

Eventually our turn came to load and, as described, we backed under the chutes to get our load. Once we had weighed and paid, we pulled off to the side in a less travelled area to eat our lunch. We couldn't take long as the long trip home awaited us and we preferred not to drive in the dark. The lunch bag was opened and we saw sandwiches on fresh-baked homemade bread along with pickled carrot sticks and, last but not least, a Japanese orange, or as we called them, Christmas oranges, because of the time of year they showed up in the stores. This was a rare find as these were only meted out on special occasions. As a family, we were lucky to acquire one of the little wooden boxes that held these oranges once during the winter season. They were much too expensive to buy any more than that, so we had to savour them very carefully. I ate my lunch and decided I would enjoy my orange on the way home.

We headed out and passed the village of Forestburg. We then headed north on the highway to the town of Killam; that's when the trouble started. We were a number of miles up the road when Dad noticed the truck pulling the steering wheel hard to one side. He stopped and got out to check and, sure enough, one of the front tires was flat.

This was a major problem. A long ways from home, a truck that couldn't travel because of a flat tire, a small boy on a winter's day in a cold truck cab, a gravel highway without a great deal of traffic, and a need to get the tire fixed in order to carry on. Dad must have had a jack and wheel wrench for I do remember him walking up the road rolling the tire before him as he headed to the next town, probably Killam. I assume at some point he was picked up by some kind soul and delivered to a garage where he could get the tire fixed. It must have been a tough decision to leave his little boy in the truck while he carried on with what he had to do to get the tire repaired. I'm sure I received some stern instructions about what I could and could not do while he was away.

But I had my orange and oh how I revelled in it. You see, these were segmented oranges and each segment could be separated from the others without breaking through to the juicy pulp. First I made the orange look like an elephant's head by peeling a section down the side of the orange to be the trunk and then pulling the top and bottom portion of peel to be the elephant's ears. His forehead part remained attached to the orange. It turned out pretty good, sort of how Dad had showed me how to do it. Then it was time to begin the most important part, delicately eating the orange segments. Firstly, one segment was carefully removed from the orange. The bottom of the segment was placed between my teeth and opened. It was then possible for me to begin to use my tongue and insert it into the orange segment and sort of separate the moist pulp from the outer membrane. This could take some time, but if you were very careful you could bring the orange pulp out in one piece and then let roll it around in your mouth for the full flavour of the pulp and juices to permeate your tasting senses. And so I spent the next while carefully savouring my orange while I awaited my father's return to remount the tire so we could again be on our way.

I don't recall much about the remounting of the tire, or who helped Dad by bringing him back. I do remember how a Japanese orange gave a little boy something to do and a great deal of pleasure while he was on his own on the side of the road for what seemed like a very long period of time.

I have had the good fortune of having many "Christmas oranges" over the years since, but I will never forget the one I had that day on our trip back from the coal mine. Without a doubt, it was the best orange I have ever tasted and is my favourite of all time.

# Chapter 20

## Prairie Fire

Prairie wool is made up of plains rough fescue, with lesser amounts of slender wheatgrass, June grass, purple oat grass, and fringed brome grass. As a child, because the land that Mom and Dad acquired through the Veterans Land Act had never been broken and had all natural vegetation on it, we had the privilege of becoming familiar with prairie wool firsthand. I well remember the times of finding meadow lark nests in the spring, well hidden in a prairie wool enclave. The reason for the "prairie wool" nomenclature was the fact that these grasses, when left alone for a couple of years, looked very much like wool covering the ground. With the bulk of previous years' growth and the tendency for plains rough fescue to stand more or less erect, only bent over with the winter's snow, prairie wool looked almost cloud-like. The grass did this without breaking or lying flat; with its very slow decomposition rate and the new growth pushing up through this covering, it made for a very soft, almost fluffy, ground cover indeed. The meadow larks and other ground-nesting species had a wonderful place for nest-building, in my estimation. Certainly the field mice, striped gophers, and other ground squirrels had a heyday as well under the cover of these grasses.

The value of these grasses were described to me as a child. According to my father, cattle and/or horses could be let out to graze on this fodder over the winter and they would come back in at springtime in better shape than when they left. I suppose this was not new information, as the great herds of buffalo had lived off this forage mixture for ages and had done very well indeed. It certainly wasn't the poor type of feed that nearly led to their demise.

Our farmyard was located a short distance north of a relatively small body of water, approximately three quarters of a mile long and perhaps one quarter mile wide. This was know by us and our neighbours as Mabey Lake. It was named after the owners of one of the early ranching empires in this area: Kemah Ranch. Harry and Seba Mabey had formed the ranch after their arrival in Alberta during the early 1900s. Kemah Ranch's deeded land surrounded the lake and it was often used as an area where they would allow the prairie wool to mature and then cut it for feed. This and other rented land was used in this venture. Lola Mabey, one of the daughters, told my father that at times during haying season as many as twenty mowers and rakes were busy putting up the prairie wool. It was cut and then raked and stacked into large loose haystacks to be used in the winter when cattle and horses could no longer graze. Before the land was broken it was easy for us to walk the areas where these stacks had been and see what seemed to be concentric circles around their perimeter. We found out that these were actually fire guards that had been ploughed around the stacks to protect them in case of fire.

Fire has been a part of nature since the beginning of time. Certainly prairie fires were not new to the plains peoples. Not only were these great fires started by lightning, but at times they were deliberately set to freshen the grass and attract the great bison herds to the renewing prairie and, sometimes, divert an enemy's plans. The Siksika prairie peoples were known as the "blackfeet" because of their feet and ankles being black from walking in fire-ravaged prairie areas.

Prairie fires were a feared occurrence once they got going in areas of prairie wool. Given a hot spring or summer's day, the heat and smoke created by the inferno had nothing to stop it from advancement; you had a formidable situation on your hands. Dry prairie wool was an ideal fuel. Once it got burning, it created its own draft. The quick flare-up with heat and smoke rising quickly and the rush of cooler air as it came in to replace the warm air made an almost furnace-like scenario.

As these fires ravaged the plains from time to time, they were doing good as well. This has been an important discovery in forested areas, too, where fire now is known to play an important role in a healthy forest canopy. On the plains, for example, fungi such as "black knot," often found on chokecherries (amongst other species), were no match for the cleansing actions of fire. Certainly the fires, for the most part, also did a good job of keeping old timber stands fresh and vibrant. Noteworthy is that early pictures of Kemah Ranch indicate a much more open landscape, with trees only in the low areas, where the water found in these impressions arrested fire advancement.

It was on one of these very warm spring days that smoke was seen south of Mabey Lake. It wasn't diminishing, either. It wasn't long until flames could be seen leaping into the air; the prairie wool was on fire. My folks soon had our neighbours dropping in to see if we could help fight the fire. A quick discussion and a plan of action developed. I remember the hustle and bustle as Mom and Dad went into action. We had an old 1952 light blue step-side Chevy half-ton. Its box was immediately filled with a couple of forty-five gallon drums and driven to the water trough, where the drums were filled with water being transferred using a five-gallon pail. My mother immediately did a sortie of the barn and outbuildings to get as many gunny sacks as possible to dip into the water and then use to beat out fire. These hemp sacks, embossed with such names as North West Feeds, United Grain Growers Feed Division, Co-op Feeds, and so on, were ideal for soaking up water and made excellent fire-smothering tools. Henry Ruste, our neighbour to the north, who was partial owner of the land, drove through the yard with a plough attached to his old John Deere tractor.

The plan included turning a furrow a fair distance from the approaching inferno and then back-firing at this point so that both fires burned towards each other; when they ran out of fuel they would die. Back-firing was done by starting a fire next to the freshly turned furrow on the same side as the out-of-control fire. The water and gunny sacks were used along this furrow line to be sure that no small flare-ups started on the other side of the furrow and thus thwarted the whole idea. Naturally I was along as an observer, basically because there was nowhere else for me to go at the time.

I remember the heat, the smoke, and my folks using their gunny sacks to beat out some flare-ups. It wasn't long before I noted, during their back and forth returns to the truck and water to replenish their gunny sacks, that this was no easy job or picnic we were on. Their faces were strained, glistened with sweat, and were darkened with soot. Once light-coloured shirts were now dark down the middle of their backs from the toil. Dark smears of ash were on their clothing and faces and still they kept at it.

The backfiring was successful. After a while the fire began to die down; all that eventually remained were areas that had heavier herbage. This wood continued to burn for a while. It was a matter now of keeping an eye on these to be sure that they didn't start another fire should a breeze come up.

I will never forget the prairie fire by Mabey Lake. There were lessons to be learned by a small child. It was important to see how my folks dropped everything and came to the aid of their neighbours in need. Certainly it reinforced the things they had tried to teach me all along. Fire is a great friend but a terrible enemy. Little hands and fingers having access to matches and the like should never, ever play with them. The smoke and heat of that day made sure these messages were indelibly imprinted in my mind for all time.

# Chapter 21
## The Kitchen Range

"Modern Comfort" was embossed at an angle on the door of the oven. It took up a prominent position in our farmhouse kitchen. And it was big. For a little boy barely able to see over the cooking surface, it was gigantic. There is no doubt it would have required some strong farm men to bring it into the kitchen and position it near the chimney.

The kitchen range or cook stove was the centre of any farmhouse in those early days. Its importance and magnificence are well remembered. The stove was mostly made of cast iron, heavy cast iron. The structure started with substantial sculpted feet or legs built to hold the considerable weight of the upper portion of the appliance. This included the cooking surface, which had four circular holes with covers over top of the firebox. The surface continued over the oven portion, ending at the outer edge of the water reservoir. At the back edge of this surface a panel arose, about the same height as the cooking surface from the floor. A shallow and short pair of doors with top hinges were at the top of this panel. Warming ovens, they were called, and were ideal for putting dishes that needed to be kept warm. The shelf formed by the top of these ovens held such things as cooking spices, utensils, and various cooking containers.

The round firebox lids could be removed with a special lifter that fit in a unique slot made for that purpose. A spring-like gadgetry circled the handle and made it so the handle was not too hot when lifting one of these covers. In addition, the two lids directly over the firebox were embedded in a rectangular hinged part of the cooking surface that could be raised to open wide the firebox area. This rectangular lid, along with the large circular lids, were very handy when adding firewood or starting the fire, and they served as a place to toss in any accumulated burnable garbage—sort of an early small version of an incinerator. Mind you, there was very little packaging garbage in those days as most everything was used again and again, including wax paper, brown paper and the like, and certainly because most of our food was homegrown, very little packaging ever reached the house.

The firebox was positioned under the firebox lids. This is where the fire was built, using kindling to start and gradually making a larger fire by making use of the accumulated coals in order to ignite larger pieces of wood. At the bottom of the firebox was the grate system. This was made up of three pieces of cast iron, looking sort of like a pipe with fingers. The ends of the grates were cradled in another piece cast with indentations specifically manufactured to accept the grate ends. One end of the grate was cast with a square configuration so as to mesh with a wrench that came with the stove. A few back and forth motions with the wrench on the grates and they would be cleaned of all ash. In our case there was no need for a wake-up call for the household in the morning as Dad moved the grates back and forth. The resulting soft grumble of cast against cast cushioned only by the wood ash gave a clear indication that morning had arrived. Sometimes wood was supplemented with coal, especially in the cold overnights of winter, to "hold" the fire so that our house continued receiving warmth during the late nights and very early mornings. With the use of coal also came the inevitable problem of clinkers. These hard volcanic-type lumps, formed from melted non-combustibles, were a job for the poker that hung on the back of the range. This tool was approximately two-and-a-half feet long, made from a piece of five-sixteenth-inch iron rod. One end had a circular loop that fit nicely into the palm of your hand, the other had a two-inch section bent ninety degrees from the main shaft; this was usually tapered, flat at one end and gradually increasing to the full round size at the bend. This was the working end of the poker; it could be used to stir the fire, move a log into a better burning position, or help remove the coal clinkers when needed. The ash fell into the ash can strategically located below the firebox to catch the spent coal and wood ash. When this ash can was full it had to be emptied and usually ended up on an ash pile beside the burn barrel one hundred yards south of the house.

This whole firebox area of the range was lined with firebrick to keep the heat where you wanted it. Naturally this added substantially to the weight of the overall structure. As mentioned, the stovetop with the covered holes over the firebox continued on over the oven area and then stopped after covering a copper water vessel at the far end from the firebox. This was filled with soft water from the cistern in the basement and was kept full at all times. All water was hauled by hand and by bucket. Our cistern had been constructed during the development of the basement of the house. It was two feet deeper than the surrounding basement floor, was approximately six feet deep, and a rectangular sixteen by twenty feet. This left it nearly four feet below the first floor joist. A floor/lid had been constructed over it to serve as a cover, with an entrance for the inlet pipe in the most northeasterly corner. This pipe was connected to the eve-trough system of the house and every time it rained the cistern received fresh soft water. If it rained too much it was simple to turn the pipe to drain onto the ground outside. In my recollection, it seemed this was not a common occurrence; actually, we were quite often looking for rain. I don't recall, however, ever hearing that we were out of water in the cistern; I do know that my parents knew how to conserve water. After all, it was back-breaking work keeping everything full and running.

The hand pump on the cistern was used to fill pails that were then lugged up the stairs to the main floor and dumped into the reservoir; it was our hot water source. This reservoir was a rectangular vessel and its cover was of the same configuration.

The whole heating surface of the kitchen stove could be used to cook things. Adjusting the damper on the side of the firebox or the second damper inside the chimney leaving the range could control how quickly the fuel burned and thus how hot the fire burned. It didn't take the lady of the house long to figure out where the warm and hot spots were on her range. I suppose you could call them the range's "sweet spots."

Of course the oven was an extremely important part of the stove. It sat next to the firebox and had the hot gases from the fire channeled around it on their way to the chimney. Ours was a modern range, because it actually had a round thermometer on the door that gave an indication of the interior temperature of the oven. And oh, the treasures that came from within its enclosure. I can still smell the loaves of bread, the buns, the three-cornered buns made for special occasions, and of course all the specialty breads and loaves. Not to mention all the pies that emerged from its cavernous interior, including apple, strawberry, rhubarb, and saskatoon, to mention but a few. Meat dishes including such culinary delights as casseroles, ribs, chicken, and of course the roast goose and turkey that inevitably showed up on special occasions like Thanksgiving, Christmas, and New Year's, and so many more dishes too numerous to mention also emerged from its interior.

The range's importance didn't end with food preparation. Mondays were wash day and the clothes were hung outside on the clothesline. In the colder winter months they dripped dry until they froze stiff. I remember my mother bringing the clothes in off the line. It was sort of like stacking dried fish—big ones, mind you. The stack of stiff clothes would stand at attention for a short period along the wall until the they began to collapse with the warmth of the house interior and then would be hung on a clothes horse near the open oven door of the kitchen stove. The clothes horse was made of wood with a number of round dowels inserted into wooden legs and arms, an apparatus that could be folded up and stored when not in use. An hour or two beside the cook stove and the clothes were dry and ready to be folded. It was so interesting watching little whiffs of steam rising from the once stiff and frozen garments; there was no need for a humidifier during those days. Often the windows were covered with ice, especially in the bottom corners where the humidity had come against the glass surface and condensed. The frost patterns on the windows were also amazing; I'm sure they have inspired many decorative ideas over the years. No two areas of the window were ever exactly the same. No doubt many fancy works of art have been based on these beautiful frosty etchings.

Keeping in mind this range in our farming kitchen reminds me of another important job that was part of its existence. Many times when I would come into the kitchen I would find a cold member of our various livestock in a box on the oven door, being revived by my mother. I've seen wet baby chicks, ducks, turkeys, very small pink piglets, and newborn calves. There was nothing like the warmth of seasoned firewood to bring a chilled youngster back to life. Extremely cold winter evenings were made more bearable as well by the warmth provided. As a small boy I remember my brother and I standing near the open oven to warm up after an exhilarating playtime outside. On extremely cold nights, chairs were pulled up around the open oven and feet were placed on the outside edge of the open oven door. It didn't take long for the feet to start tingling as the socks took in the radiated heat. Some reading by the adults of the house also took place as they sat in the warmth; eyes became heavy and a few winks were snatched under the spell of well-being cast by the oven of the kitchen stove.

As we had to leave the comfort of the kitchen and head to bed in the evening, a brick that had been warming in the oven would be wrapped in a towel and accompany you to your bed. How nice it was to have this warmth when crawling into the frigid sheets and blankets.

Of all my memories of the cooking range and its important role in our farmhouse, I will never forget the night when the stovepipe started to turn red and my folks began to cry "chimney fire." They had obviously seen this before, and had also known of farmhouses that had been destroyed because of a fire started in the chimney. As firewood and coal burned, sometimes the fire wasn't hot enough to burn all combustibles and the resulting smoke was laden with unburned particles that settled and attached themselves to the interior of the stovepipes and the brick chimney. I recall seeing my father up on the roof with a piece of very heavy logging chain attached to the farm lariat, dropping it in and pulling it up and down as he worked around the perimeter of the chimney. This rubbed and loosened the attached soot and creosote and it dropped to the bottom of the chimney, where it could be removed through a small door expressly for that purpose.

The evening of the fire, it didn't take long for my parents to grab some towels and oven mitts and the stovepipe was knocked apart in short order. My father immediately threw the lengths of pipe out in the snow. After all, with the fire inside them, they were red hot and needed to be cooled. Swiftly all the pipes were out there and my father began immediately to clean the insides. Keep in mind that the cook stove was still belting out smoke and the only place for it to go was in the kitchen. Soon the doors and windows were open, and with a great deal of haste Mom and Dad had the stovepipes cleaned and back in place. What was one of the most exciting days I can remember I'm sure was not recalled with the same fondness by my parents. They were thankful that an all-out fire had been avoided, but the cleanup of soot and debris was no pleasure; it was everywhere.

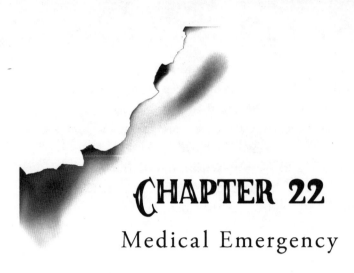

# CHAPTER 22
## Medical Emergency

Intussusception: *Kids Health* describes it as "a serious medical condition where the small intestine telescopes into the large. It is the most serious abdominal condition in children under two years old. When this "telescoping" happens, the flow of fluids and food through the bowel can become blocked, the intestine can swell and bleed, and the blood supply to the affected part of the intestine can get cut off. Eventually, this can cause part of the bowel to die. Babies and children with intussusception have intense abdominal pain, which often begins suddenly and causes the child to draw the knees up toward the chest. The pain often makes the child cry very loudly. As it eases, the child may stop crying for a while and may seem to feel better. The pain usually comes and goes like this, but can become very strong when it returns." In my case the pain was so intense that it actually knocked me out and I would go limp, only to reawaken a few minutes later and repeat the proceedings. I do not remember any of this, of course, as I was only five and half months old, but I do remember talking with my mother and father about this time and certainly I have had a large incision scar on my abdomen to prove that this did occur.

Luckily Mom was a nurse, and a good one, I've learned from a number of sources over the years. It was said that she had extraordinary observation skills. These skills were legendary and there were many individuals that lived much longer lives because of her surveillance and the subsequent information passed on to attending physicians. That was certainly the case in my situation, as it was Mom's observation of what was happening to me that led her to suggest to the doctor, Dr. Bradley, that she thought I might have intussusception. Knowing my mother, she would have suggested this without arrogance or being forceful in any manner; that was not her way. She would have quietly suggested the diagnose, shyly but firmly. She would have "all her cards in a row," as it were, and would have known what she was talking about.

What to do? It was decision time for this young couple with a three-and-a-half-year-old boy and this five-month-old. Mom would have been twenty-nine at the time and my father thirty. I'm sure they agonized about what to do next with this situation thrust upon them. They had my grandparents, of course, and a call was put in to my uncle Allen, Mom's younger brother, who was interning at Shaughnessy Hospital in Vancouver. Naturally their close friends were notified, including the ministers of the church. I'm sure a great deal of prayer was uttered on my behalf during those crucial hours. Many factors were considered; time was of the essence. There was my folks' old car that was at least ten years old. I believe it may have been one of the later Ford model Ts. It wasn't the fastest vehicle in the world and Edmonton was a long way to go. The roads were gravel east of Tofield, some 150 kilometres west of our hometown and three-quarters distance to Edmonton. This was our closest major centre; it took approximately three to four hours to get there on a good day and then more to the hospital after reaching the outskirts.

I was taken to the local cottage hospital and evaluated by Dr. Bradley, who agreed that I probably had intussusception. Unfortunately, as a young doctor, he had not done the operation of correction for this medical emergency, but had been in the operating theatre as a student when one had been performed. With all things considered and knowing time was of the essence, my folks asked Dr. Bradley to do the operation and he consented. He immediately checked out of his clinic and sat down to go over every detail in his medical books and journals, to be sure that he was as familiar with the procedure as possible. He quite possibly also consulted with some of his colleagues to be sure that he was well-prepared. The time was set and I was whisked into the little operating room of the Wainwright and District Municipal Hospital. Everything went well, I'm here to say. Mom related that the next time she saw me was when I was ushered into the recovery room. I was wrapped in cotton baton with only my face showing. This was used to keep me from going into shock after such an abdominal invasion.

In cases like this the protocol of the day was that a special nurse was assigned to look after the patient. Dr. Bradley asked my mother if she would "special" me. Naturally she agreed, perhaps because she, like any normal mother, would have been unable to set aside the plight of her baby and being a nurse she could personally watch over my recovery. It couldn't have been easy, however; in the crucial first few days the outcome was still in the balance. I'm sure my father would have also spent some difficult days with his mind preoccupied with his small family, keeping the farm running, and the difficulties they were experiencing. Naturally the extended family filled in where needed. I'm assuming my older brother would have ended up at my grandparents', which was probably a bonus for him, knowing what softies grandparents can be. In any case, I never looked back. My recovery was quick and it wasn't long until I was doing well again, doing my share of complaining and crying, I'm sure. This probably sounded like music to my parents' ears after the ordeal they had been through.

All went well until it was noticed that the thin tough layer of tissue called fascia had obviously separated and a hernia developed where my incision was located. The stitching of this layer had let go, and was thus causing the problem. As you can imagine, the tissue of a five-month-old is not only tender but, size-wise, very difficult to work on. It was monitored over the next while and a decision was made to re-operate once I had had my second birthday. By then it was expected that the tissue would be firmer and the hernia could be repaired.

I honestly don't remember much of this adventure, but again I was taken to the local hospital and once again Dr. Bradley had me for a patient. The repair went well, I'm told, and I spent a few days recovering from this second major operation. I'm sure most of us don't remember much from when we are two years old. I don't think there was any snow on the ground, so I believe the operation would have taken place some time in the spring months, perhaps April or May, which would have made me nearly two-and-a-half. At any rate, there are two things I do remember from that time. I'm sure these memories have been supplemented with photos and stories told; however, I do remember a couple of things clearly. Firstly, Mom had been busy while I had been hospitalized and she had sewn both my brother and I cute little sailor suits. They were navy blue with wide seamen's collars complete with white stripes across the bottom edge. The outfit was completed with real-looking sailor hats. If they had been given to us as grown men we could have easily joined the Royal Canadian Navy; they looked that good. Secondly, I do remember walking out onto the steps of the then-new municipal hospital and seeing the car and my brother in it. It had been days since we had seen each other; the anticipation was breathtaking. With all my strength, I hollered, "Hi Tent!" His name is Kent and it seemed he was just as glad to see me.

Fortunately the stitching held this time; I never again was hospitalized for this condition and I call him Kent now.

## Keith Erwin Brower

# CHAPTER 23
## My Grandparent's Farmstead

I remember Grandfather and Grandmother Dixon's farmstead fairly well. I was rather young when I was first introduced to it. Probably there were many times prior to my forming conscious long-term remembrances that I stayed in the house while my parents were busy doing things that required their attention. There were no end of things that needed to be done on a farm, especially when the only work force was my mother and father.

Grandpa and Grandma had a nicely painted white two-story house with an attached porch on its east side. The front door had a veranda with vines crawling up the supporting posts. They were yellow clematis, one of Grandma's favourites. In addition, there were pansies and stocks, bleeding hearts, columbines, peonies, and roses, amongst many others, and always lots of delphinium. Grandma loved delphinium. Their yard was always nicely kept and one of the most impressive things for a very young fellow like me was the push lawn mower they had. It had a reel cutter head that passed over a knife to cut the grass. The whirling mechanism from the ground-driven wheels no doubt caught my attention. Following along behind was the lawn roller that the whole contraption rested on. I guess anything with wheels was irresistible and small boys were allowed to push the mower; however, the grass was rather safe because it took a great deal of effort to do anything with the machine.

Between the farmhouse and the road there was a very sizeable sandy loam garden patch. I particularly remember that on the west side of this garden was a largish patch of asparagus. Come spring, it delivered an abundance of succulent green shoots; combined with butter and salt and pepper, there was a feast at hand. I learned to love this garden cultivar in those early years.

My grandparents must also have had a cistern in their basement, for at the east end of the kitchen counter was a sink. A hand pump was mounted at its side that brought water up from the cistern below. It was at the kitchen sink that everyone washed up for meals, including us boys. My grandfather washed his face by cupping his hands under the water in the sink and bringing it to his face in what seemed to me like a great splashing and rinsing event. Of course there was a need for an abundance of water because the spring and fall work often left the faces of those working the soil very dusty and dirty indeed. Especially in the spring, sometimes it appeared as though my grandfather had changed his ethnicity, with only the whites of his eyes and his white teeth showing when he smiled. Other than that he was completely black with dust.

I do remember being in the basement, which was entered by lifting a trap door that had a large ring for grasping and hinges on one side. This trap door was located in the main kitchen area away from the table and chairs for easy access. The door itself was lifted by the ring and laid over to the hinge side, which allowed you to descend the rather steepish stairs into a dimly lit, earthy-smelling space in the ground. Basements in that day were not like the lower floor of today, but more like an enlarged hole in the ground where shelving and cupboards stored preserves, garden produce, garden tools, and so on. An outdoor entrance with a "door laying flat" type of covering over another set of stairs let you in from the exterior of the house to the basement as well. This was a very useful entrance for bringing in and taking out all the different things stored in the lower level. It was not so good at keeping out the small "livestock" that enjoyed a warm place for the winter and any food that might be available. Mice were always around and very much enjoyed the dirt walls and areas in this basement that gave them lots of places to tunnel and raise families. Naturally, with these little critters around, the farm cat was an important member of the family. Grandpa and Grandma had several over the years and they usually kept the mouse population at a manageable level. Many times as young children we would watch as one of these cats would catch a mouse but not kill it. They would play with the unfortunate creature by grabbing it and then letting it go and re-catching it again and again. Normally this would carry on for a while until the mouse was nearly lifeless. At that point the cat would crunch the mouse and give it a few quick bites head first, tail last, and swallow the whole thing. I have also seen a cat fool around too long with its prey and the little rodent escape.

Of all the interesting objects that could be found in the basement, the most interesting to me was the "dumbwaiter." This was a device that resembled a modern building elevator. A shaft ran from the basement up to the kitchen and rose behind a cupboard door. A rope attached to the dumbwaiter car ran up through a fixed pulley and back to a counter-weight that weighed just slightly less than the loaded moveable cupboard. This rope was near the open cupboard door; by pulling it up or down, the car was raised to the kitchen or send down to the basement. Of course, the whole idea behind this was the coolness of the lower earthen cavity. Certainly there hadn't been electricity on the local farms until the early 1950s, so naturally there were no refrigerators. And even if a fridge was available there was the small matter of being able to afford it. In those times credit wasn't readily available, so people saved up for a long time and if more important needs didn't require the cash, then a purchase could be made. In the meantime the dumbwaiter was one way the housewives of the day were able to keep perishables a little longer. Things needing to be kept cooler were loaded onto the dumbwaiter and the rope pulled upwards, thus lowering the moveable cupboard in the shaft until it was at the bottom.

Things needed to be well-covered in order that the food not be eaten and messed on by the few mice that had escaped the watchful eye of the cat. I remember Grandma saying that once a mouse caught a ride up the dumbwaiter and when she opened that cupboard door, out jumped the mouse. I'm sure the resident cat would have heard the ruckus and been on the job quickly. Must have been a real thrill for the little rodent riding the car up the shaft, but I'm quite sure it didn't end well.

The other interesting way my grandparents preserved food and vegetables for the winter was in the root cellar and part ice house a little ways east of the house. A set of stairs led to an underground dug-out room where the ice and vegetables were kept. Ice was used in the icebox, the forerunner of the refrigerator. This was made up of a wooden box that was covered with galvanized steel painted white and, in this case, housed along the wall in the kitchen. With a block of ice placed in this insulated box, food could be stored here as well and because of its coolness the icebox was able to help keep food from spoiling. The ice block used for this appliance came from the below-ground storage area where it had been placed during the past winter.

When I recall my grandparents' farm home and recollect the arrival of electricity in rural Alberta, the icebox was being replaced with refrigerators. Not that long before my time, however, the farmers in the area, including my parents and grandparents, would procure ice for summer use during the winter months. This involved finding a decent body of water that was well frozen over. Once the ice had reached a good depth, around twelve inches, the work began. Using large ice saws, ice tongs, axes, and brute strength, people would cut and load blocks of ice onto a horse-drawn sleigh. These blocks were then delivered to the icehouse, carried down the stairs, and stacked. Sawdust was used between them for insulation, which slowed the melting process during the hotter summer months. This then provided the cooling source for the icebox. In addition it cooled the room of the root cellar, providing good crisp vegetables all season long, until the next growing season delivered its bounty.

Another delicious use of the ice was when ice cream was made. This was the real authentic thing, with real cream and home-grown eggs and ingredients. There was the chipping of the ice and the cranking of the ice cream mixer's handle (which seemed to take forever). Eventually, as you were turning, it began to take more and more effort and finally for a small boy it was almost impossible to crank the handle. It was at this point that an adult usually took over and finished the job. And then the chance to savour some of this decadent delight was at hand.

Naturally the pioneers of the district used what was available to them in their environment. Things like ice, for example, were used to make life easier by cooling and preserving food for our parents and grand-parents. But not all ice sources were welcome or equal. The most interesting story I recall is the one that points out the resiliency of these early Albertans. It's about the ice required for making ice cream and an unusual source. This source was the ice found all over the ground after a summer hail storm. Can you imagine a time when crop insurance did not exist and you as a family have just watched the year's field-crops and garden being wiped out in a matter of minutes in one of these severe storms? Rather than wallow in self-pity, the ice cream maker was brought out and a treat for all to enjoy was produced.

My parents and grandparents were made of stern stuff, the kind of fabric that when woven with many others like them has made this such a great country of achievement, progress, and one that I have always thought I was very lucky to live in.

# CHAPTER 24
## The Cunning Coyote

They have been around for eons, and certainly figured very early in my memory: the "wild dog of the parklands and plains," the cunning coyote. It seems like we have had a love/hate relationship over the years. It was always fascinating to watch one of these canines as they hunted rodents. Still as a statue, one paw raised, eyes focused on the ground, ears erect—and then the pounce. Both paws hoping to trap the day's breakfast between them and the ground. One quick nip and the rodent's career was over and it had become a tasty delight. It only took seconds, but the concentration part could last a fair while. This quest for food included not only mice, but also grasshoppers, small birds, birds' eggs, wild fruit, and, in fact, practically any creature or soft dainty that appeared edible. Naturally this was applauded by all who had watched the event, as mice and the like were also an enemy of the farm families trying to make a living off the land. Especially mice—they could get into feedbags and chop bins, chew up harness, and be a general nuisance with their gnawing and their excrement making a mess, rendering foodstuffs, tools, and harness unusable.

However, included with this array of edibles were the farm's fowl that my folks raised for eggs and meat. The wily and crafty coyote also enjoyed a chicken dinner when one of the girls wandered off from the main flock or found a hole in the chicken-pen fence. This sly scavenger made a quick end to the life of one of these unfortunate domestic birds, and a warm, tasty meal was at hand. No doubt during the spring season a family of pups was probably involved, either being taught to hunt or certainly sharing in whatever could be brought to the den. The unfortunate part was that one successful hunt near the farmyard often became habit-forming and a great deal of vigilance was required on my family's behalf to have some hens left for the winter's egg supply. And coyotes liked turkey, ducks, and geese just as well. So the relationship of the scavenger and my family would change. Now the hunter became the hunted and whenever they were seen around the farmyard they were fair game.

During the winter months when the coyote's coat was prime and they became a pest, Dad would get cyanide "guns" from the municipal district field-man. They were specifically designed to take out problem coyotes, and that they did. Although I never saw one, I do know that they were set with a bait such as meat and, when pulled on, shot cyanide into the coyote's mouth and thus disposed of it very quickly and humanely. During the harsh winter months, when the coyotes' coats were prime, Dad would tan a few of them and Mom would use the fur to line the interiors and edges of our parkas and hoods. The fine inner hairs, combined with the darker guard hairs, made for very warm faces when we were exposed to cold winter winds. I'm sure some also entered the fur market as there was always someone in the area gathering up pelts of various animals, including coyotes, beaver, and rabbits, for sale to the furriers. A few extra dollars could be made and the problem coyotes were no longer in the farmyard causing grief.

Often, my brother and I would team up with the Miller boys, who were our age, and head to Mabey Lake to clean off a patch of the accumulated snow to skate. Usually the cleaning took most of the evening with a short time of skating to follow. I can still see the small dark patch surrounded by the sparkling snow radiant under the moonlight. By the time we had cleaned off this petite rink, we were warm enough to put on any old skates we had found and a short version of a hockey game began. We were rich indeed, as we skated around the little area, with "diamonds" sparkling in the moonlight all over the snow, the most brilliant ones found where the moon reflected off the snow as its light made a path all across the lake.

One of the drawbacks of these times involved the inherited need of coyotes to howl at the moon. Often on these types of evenings a great chorus would start. First one, then another, and soon a great choir could be heard from all directions: a wild choir singing their melancholy melodies. For small boys, this was enough to send chills up our spines and the hair on the back of our necks was raised in a stance of high alert. After all, we had all heard stories of wolves and how they were bloodthirsty; their prairie cousin couldn't be much different in our minds. In this case the skates were off in record time, and four little boys with tied skates over their shoulders, one scoop shovel in hand, having made a few practice swings with hockey sticks and the shovel in a protective mode, were off to the farmhouse as quick as their little legs could carry them. Great vigilance was practiced on all sides, and especially behind, to be sure we weren't the next on the list of those to be devoured. It couldn't have taken us too long to cover the less-than-a-half-mile to the farmhouse.

On one occasion I remember my mother describing our arrival on the porch steps. Four little boys, eyes as wide as saucers, all babbling at once about the chorus that we had heard and how we had barely escaped with our lives. Later we found out that the hired man had started the whole episode by howling like a coyote. This encouraged the real group to join in and show him how it should sound. Either way, the effect on us was very real; the eerie howl of the lonesome coyote still does get my attention, and when I'm out alone can still make me shiver.

Keith Erwin Brower

One of the important appointments of the coyote is as one of nature's scavengers. They join the magpie, ravens in the winter, and the common crow in the summer, along with any number of smaller gnawing animals, to clean up the carcasses left by unfortunate mammals and birds who have lost their lives for any number of reasons. Many times as a child I witnessed the tramped-down snow and blood-coloured area around a downed deer or other mammal that was being stripped and eaten by the coyotes and their group. It wasn't long until all that remained were a few major bones and even these were carried away to be gnawed on in some secluded area. A few days and a fresh snow was all it took to erase the memory of some unfortunate animal.

My most memorable encounter with this wild canine occurred after the loss of one of our prize dairy cows. The old adage that, "when you have livestock, you will have dead stock" was played out on one of our prized milkers. Any cow loss was huge in those years, and so was this loss. Added to that, it was a deep snow year. As this was in the early days of the farm, the equipment was rather primitive; with an old tractor that was rarely started in these colder months, with only two-wheel drive without the non-slip-type differentials available or front wheel assist, the unfortunate creature was dragged to the end of the lane out towards the west, and that's where she remained.

The scavengers descended. During the day it was the raucous voices of squabbling magpies that filled the air near the end of the lane. No manners here. Their ability to smell a meal brought everyone in the region, it seemed. At night, out of the shadows crept the coyotes. A feast such as this was just too much to pass up, even if it was close to the farmyard buildings.

No doubt there was some squabbling amongst them but I don't remember hearing about it. At any rate, my brother and I decided to venture out one night to see what was going on. We had our coal oil lanterns, the ones that my father had given us to use on the sleigh boxes he had built for us. We had been carefully instructed on how to light these by lifting the globe with the handle provided and striking a match and applying it to the wick, then carefully blowing the match out and making sure that it was cool before we discarded it. Then the globe was lowered and the wick adjusted to give a bright light without smoking. Too much wick and the resulting smoke blackened the globe with soot and dimmed the light. These were lanterns especially designed to be able to work even in a wind without blowing out. They stood about a foot high and had an upper handle nearly the same length that was handy for carrying or hanging from a nail in the barn to shed light all around.

So we removed the lanterns from our sleigh boxes, lit them, and slowly made our way to the end of the laneway. What we encountered is still embedded in my mind. I can still see the movement of grey objects and the reflection of eight pairs of eyes glowing in the dark. There was a drifting of bodies as they melted into the darkness and, as I recall, they weren't necessarily in a big hurry. In fact, I distinctly remember one member of the group, who had no doubt had his or her fill, curled up in a small indentation, nose under tail, enjoying a full stomach and the warmth provided by the fur coat that nature had supplied for such a cold winter evening. Naturally we were thrilled and chilled by our discovery at the same time. Needless to say, we didn't hang around long, but slowly backed our way down the lane, lanterns in front of us, until we had reached what we thought was a safe distance, where we turned and ran to the security of the glow of the farmyard light. Looking back, I still shudder a wee bit thinking of the encounter of that evening, and the thrill and chill two small boys had at the hand of some local scavengers.

Even now in the evenings it's not unusual to hear coyotes howl. If there is a change of weather in the works these interesting creatures have an inherent need to let everyone know what's coming. There is something about a weather system that is naturally detected by them. And it's reassuring, in a way, to realize that the descendants of our childhood encounter are still trotting on the fringes of the farmyard, still looking to do what their forefathers have done for generations. As I lie in bed and listen to them "sing," the thought that all is well with the world rises to the forefront of my thoughts, and I drift off to sleep.

# CHAPTER 25
## The Blue Chevy

Going to town was not something that my family did often. In those early years our transportation was limited. I do remember an old rust-coloured Dodge—or was it a Fargo?—truck. I believe it was preceded by a Ford Model T, but this vehicle had vanished from the scene by the time I was of any age. The old Dodge was the vehicle for all conveyance, whether human, animal, or goods. No doubt it had seen most of its useful life before my parents got ahold of it; as to its condition or where it had spent its best years I cannot testify, mainly because I was very young and knew very little about vehicles. The chassis of the old girl still exists on the farm today as the undercarriage for the wagon that has a rack on it used for hay rides and yard clean-up. One important thing about the "wagon" are the wheel studs. My father taught me about them and how on the right-hand side of the wagon the studs have left-hand threads and on the left side the wheel stud bolts have right-hand threads. This, of course, was very important to know if you were to change a tire—which way to apply pressure on the tire wrench in order to loosen the wheel bolt studs and remove the tire and rim. I'm sure the bolts were probably good steel so it was unlikely that you would break a stud bolt, but slipped wrenches and barked knuckles are no pleasure either. This was mostly a Chrysler idiosyncrasy (with a few other manufacturers joining in for a few years) present on vehicles until the 1970s. Dad said it was to help the rotating wheels by having the wheel studs remain tight and the wheels not come off.

I'm assuming the Dodge brothers, Chrysler, built the truck in the late 1940s. It was no luxury unit, for sure. Most of these early vehicles were built for work; the controls and cab were more for function than for pleasure. It was not uncommon to be able to see the ground through such things as the floor handle to put on the parking brake. Holes such as this existed in the "firewall" between the engine compartment and the cab itself, if indeed there was a firewall. Heaters were in their infancy, providing very little heat. Certainly the winter weather was not completely shut out, and it was not unusual to share the cab with small furry friends who enjoyed free range through all of these openings. Spill a small amount of grain or a few crumbs on the floor and it seems the word went out that a banquet was at hand. It was never fun for my parents, especially my mother, to drive with some of these guests roaming around.

The cabs on these early trucks were not roomy, either. Put a young couple and two young boys across the seat and it was pretty much filled. Seat belts and car seats were non-existent. Speeds achieved were at a minimum as well but probably were equal to the condition of the roads of the day. At some point the old girl gave up; I never heard the reason but I'm guessing she was worn out. At any rate, a replacement showed up: the one that is most embedded in my early mind, a robin's-egg blue 1952 Chevy. She was a step-side, with round rearview mirrors and a six-foot box with alternating wood and metal slates in the box bed. The tailgate had a hook on each side with a span of chain equal to the length needed to lower the tailgate so that it was level with the box bed. These chains had a piece of rubberized hose threaded over them, which probably kept them from tangling and pinching fingers. When in the down position the tailgate lengthened the bed by eighteen inches or so, which was handy when carrying materials longer than six feet. As with all trucks of the day, she had a two-wheel rear drive undercarriage, with a simple rear-end and a four-on-the-floor gearshift.

It was in this truck that I had some of my most memorable rides. Naturally it was used to deliver milk cans to the gate, where an insulated van-type truck picked up the milk that had been loaded on a stand built the same height as the floor of this van. For a period of time the milk and cream we produced was sent by this truck to Viking to the Northern Alberta Dairy Pool depot, where it was used to make dairy products for urban consumers. The blue Chevy was also used to go out in the pasture and retrieve calves that had been born there. Throw the calf in the back with the boys and the old cow would follow along behind the truck, smelling and hearing her offspring call. The odd time a rambunctious youngster, hearing its mother's deep-throated rumble or bawl, would jump over the side of the box. We boys would sound the alarm and Dad would be out of the cab in a flash, grabbing the little guy again before it could get away. Naturally the box could also haul the small square bales that were just beginning to gain popularity at that time. It was used to move gravel, dirt, and perhaps some manure over the course of a year. It also could be driven out in the field to be loaded with the Alberta field stone and roots that were part of the agony of developing new land. The sides of the top rim of the box also had three square evenly spaced two-and-a-quarter-inch square holes that were capable of holding wooden stakes. These were built into a mini corral system right on the truck called stock-racks. The truck was then able to be used to deliver hogs, sheep, or cattle to market. The racks could be lifted out of the stake holes and stored for future use after the job at hand was finished. Generally the farm pickup became an invaluable member of any farming family and the little blue Chevy was no exception.

My parents were people of faith. Being farmers and so close to the land did not make this difficult. It seemed only natural to realize that a beautiful sunrise or sunset and indeed all the multitude and complexities of nature must have a master planner. Dad was great at appreciating nature in all its fullness. He fed the birds, he took us for walks and showed us nests of birds and insects, he paused to watch the flight of birds in the spring and fall, and he could mimic the sounds of most wetland and song birds. His rendition was so realistic that many of the birds would respond and in some cases come to see who the new guy on the block was. All of this contact strengthened his faith and he shared it with his children. Sundays were for most part set aside as a day to go to worship, do only necessary chores, and relax and rest to get ready for another hard week ahead. With all the physical work in those early days, a Sunday afternoon nap was not only a luxury but a necessity. It was after this afternoon nap that on many Sundays we were taken on a nature walk in the back pasture.

Getting to church on Sunday was accomplished by everyone piling into the blue Chevy half-ton and heading to town. This worked fine when my brother and I were the only children accompanying my folks. As the family grew, finding enough room became an issue. In the winter months we all piled in and sat one or two deep. With the lack of heat in the old truck, packing in had a warmth advantage. It is the summer trips to town and church that are most embedded in my memory, however. Dad had come across an old car seat, probably at an auto wrecker's. I distinctly remember it was grey with dark grey narrow strips running from front to back. It was only the bottom with no back and rested on some boards in the front of the truck box. Come Sunday morning the old seat was thrown into the truck and Kent and I rode to town with the wind blowing and parting our hair. It was a thrill to have this feeling of freedom and exhilaration as we were driven in the fresh air and it passed over our heads and outstretched hands. Of course this sort of travel has been made illegal in today's world, but I will never forget the rush of standing on the seat with our heads over the truck cab, our hair pasted back by the wind, and our eyes squinting and watering as we travelled to Sunday services in the old blue Chevy.

Chronicles of the Glen

# Chapter 26
## The Door-to-Door Salesman

W.T. (Will) Rawleigh was the name of an eighteen-year-old man who, against his father's wishes, borrowed a blind horse, bought a buggy, and started selling medicines door to door. Within three years he had paid off the mortgage, bought a new rig and horses, paid for all his freight and wholesale goods, built and furnished a house, and had a net value of over five thousand dollars (a small fortune in 1892). His company eventually developed seven divisions: medicinal, nutrition, gourmet foods, cleaning products, personal hygiene, and animal products. At some time or other during my growing-up years our family probably used products from every one of these different categories.

Even in the 1950s my parents did not travel to town much. Certainly most of their treks were connected with commerce. This consisted of only goods that could not be grown, hewn from patches of bush, cultivated, or raised. This included the growing of your own poultry, for example. Setting hens were placed on fertile eggs of turkeys, chickens, ducks, and even geese. The mating pairs were encouraged to set on their own eggs as well. Fencing materials were cut and treated with bluestone to make them last. Barbed wire was rolled up and used again and again. Cows calved and hogs had piglets. Dad usually had a couple of sheep running with the cattle as well. For our own household use these animals were butchered and cut up on the farm. Large gardens were planted; we grew lots of basic vegetables. And the sweetener for most uses, such as cooking and canning, came from the two or three hives of bees my father kept.

Over those early years, patches of strawberries, raspberries, currents, and gooseberries were planted and grown. Planting stock was usually acquired from neighbours; plants were traded back and forth, everything from flowers to rhubarb. These were days of very little cash and a whole lot of hard work and ingenuity.

So it was into this atmosphere across rural Alberta and in fact across rural North America (and certainly on our farm) that the Rawleigh man filled a void. It was a great day when he showed up at our place, especially for the kids. I've noted that the younger set hasn't changed much over the years, still asking for things that they can't have and insisting they need them. Sounds very familiar. That's how it was around the table when he arrived, with a polite knock on the door, never a heads-up phone call to be sure you were home. I'm sure he rarely showed up to a farm home when the lady of the house didn't answer the door; after all, most farm wives worked alongside their husbands on the farms and it was the rare occasion when farm wives worked at other occupations.

And so the visit began. There was the small talk of the happenings in the district, who was getting married or buried, the latest news on additions to the district, and so on. You had to be a good visitor to be a good Rawleigh man. And then it was time to open the Raweigh kit, which as I recall was a marvellous satchel or suitcase that when opened had a set of cascading shelves laden with all kinds of products of all different shapes and sizes. I well remember the shiny metal infrastructure that supported the shelving. It was a series of short fine metal pieces riveted together in such a manner as to fold up when the case was closed. There was a soapy, medicinal aroma that wafted through our kitchen after the opening, one we children drank in as part of a wonderful event happening in our very own house.

There were the old basics that we had used since I could remember: Rawleigh's Liniment, Medicated Ointment and Antiseptic Salve, Penetrating Rub, and Mustard Compound, amongst many. Naturally some of these were put to good use, when you consider the hard work performed by my folks on the farm. A good rub of liniment on a sore back did wonders. Our cuts and bruises were liberally applied with medicated ointments and antiseptic salve, then wrapped with strips of old white sheets that had been boiled to make them aseptic. Band-Aids were new and too expensive for the family at that time. We children were not the only recipients of some of these salves, as quite often they ended up on a windowsill in the barn, having been applied to one of the cows. The one that I'm sure Kent and I remember the most from that era is Mustard Compound. This was rolled out and applied on a chest that was harbouring any kind of cough or cold. It was so strong that it made your eyes water; it was very hot. An application of this substance certainly warmed you up, and it was the kind of product that you tried your best to fake your wellness to avoid. Just the smell of it made you feel better, or so it seemed.

Another thing that struck me about the products were the bottles, cans, and labels themselves. The salves were presented in flat tins with beautiful ornate etchings surrounding a clean, neat-looking picture of W.T. himself. Others, like liniment bottles, were long and slender with paper labels showing a smart horse-drawn buggy with a well-dressed driver in the apex of health. There were wonderful etchings on the bottles and boxes of peppers, cinnamon, cloves, and all the other spices available. Some showed the plants the extract was from, others pastoral scenes that could make the hardest and most discerning housewives cave into purchasing.

Our Rawleigh man drove what must have been the first Volkswagen van in our area; maybe it was the first in Canada. At any rate, the cascading case was only a teaser really, for in the van there was a limitless stack of goods for replenishment, just waiting to be called for. As I recall, there were any number of shiny brochures extolling the value and the latest in each category that W.T.'s company procured or manufactured. These accompanied the case and were laid out on the kitchen table for all to view. Sometimes the salesman had to retreat to the van to retrieve some goods not in the case and it was at this time that we as small boys would re-strike the chorus of what we wanted and needed. I don't remember winning over my mother's sympathy very often; there just weren't the funds to make it possible.

Discussions took place about the virtues of many of the products. These often included the shiny brochures and the new and latest that they presented. Of course, a number of these were left with the housewife to read and review. There would be another house call in a month or so and perhaps if enough money was procured between now and then some of these new products might be tried.

Our Rawleigh man, George Foged, was a kind gentleman. He understood his clientele and their limits, only plying them with things they needed and could afford. After all, he had a family of his own and he well understood that in order to maintain his clientele he needed to treat them fairly. He lasted a long time at the job, in my recollection, so he obviously had a good grasp of the required concept.

There was another thing that kept my brother and I watching and alert to his arrival. At the end of the visit, the Rawleigh man always handed out a gift to all the children present. It was a small package of Lifesaver-like candies in a short roll, probably three or four in all. They had a silver inner wrap and a label around the exterior indicating they were W.T. Rawleigh candies. They were light green in colour and had a minty flavour; they were wonderful. If you were very careful and didn't suck them too hard and removed them from your mouth and saved them, you could make those three candies last a number of days.

Often I think of the important role folks like the Rawleigh man played for the early housewives of our region. However, times changed, as they always do, farmwives began to work in nearby communities, and shopping patterns changed. Gradually, the need for door-to-door goods delivery slowed to a trickle, and these door-to-door salesmen disappeared or retired. As I look back, however, I would love to see the old blue Volkswagen drive slowly into the yard. It would be so much fun to see the old cascading case open in front of me, but mostly I would love to place one of those mint candies in my mouth, close my eyes, and pretend I was the driver in the buggy on one of the Rawleigh labels, driving down an apple blossom lane with the sun on my back and the aroma filling my nostrils.

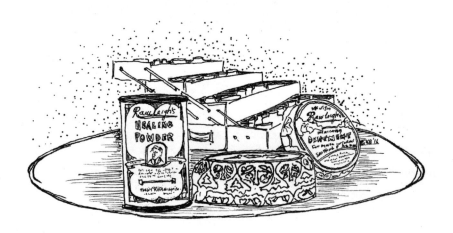

# CHAPTER 27
## Taking the Train

With clouds of steam belching and mixing with the acrid smoke of the boiler, the whole cloud clinging and encircling the engine, the huge black locomotive emerged from the darkness. The sound was deafening, with the squealing of brakes, the engine's bell ringing, the hissing of steam, and the clack of wheels passing over rail joints. The pistons and wheel-driving cranks slowed, the ground shook, and the windows in the waiting room rattled slightly as the Canadian National Super Continental made its arrival known. I will never forget its arrival. The fluted reflectors of the electric lights of the platform made an interesting pattern at the edge of their influence and they shed light on the arriving train and all the activity that was part of the happening. We had been in the waiting room for some time. The room itself had its own character, permeated with years of smoke from pipes, cigars, and cigarettes mixed with the smell of oiled floors, amongst many other odours drifting through the station. The walls carried an aged look from this onslaught as well as the general wear and tear that the public had inflicted on the building but even at that it was impressive and beautiful. Wide wooden baseboards, panelled windows with large casings, all of them an appealing dark-coloured wood, were part of the decor (designed by the Grand Trunk Pacific Railway). Massive cast iron heating radiators belched out heat and the floors sported speckled dark green commercial flooring throughout.

As you walked into the station, straight ahead was the station agent's office, where he and the telegrapher resided. There was a counter there and a grilled ticket wicket where patrons stepped up to make their ticket purchases. It seems to me there was a great deal of chatter as these tickets were bought, destinations were clarified, and luggage was parted with. The purchaser eventually ended up with one ticket copy, the station agent with one, and I'm assuming one for the train conductor. Next to the wicket was a blackboard listing the trains, their numbers and times of arrival, and a most interesting notation of "OT," which I learned in later years meant "On Time." Often enough, however, the station agent had to fill this slot in with the expected time of arrival not being "On Time." There was an impressive list on the board with two passenger trains going east and two going west every day. During holiday times such as Christmas, the Super Continental and the Panorama each ran two sections; in other words, there were four passenger trains going east and the same going west daily. These trains were usually twenty to twenty-two cars long, carrying hundreds of passengers.

Of course, to a four-year-old, all this was of limited interest. I wasn't able to read what was on the board (in fact, the counter seemed a long way up) but I found that if you stood back a bit it was much easier to see. This was not a difficult thing for me to do since as a small farm boy I was rather shy and had never seen anything or been anywhere like this.

In the waiting room were several tri-seats made of oak, substantial in design, with padded seats. A row of a couple of these were in the centre of the room with several others around the perimeter. Here a number of people sat waiting for the arrival of the train with a few smaller bags scattered around their feet. Our larger bags had been given to the station agent at the time of ticket purchase and they had been taken to the baggage room, where tags were attached with our destination stamped on them and, of course, our names. Once that had been accomplished the suitcases were loaded onto the baggage cart, a wagon whose floor was the same height as the floor of the baggage car. Each passenger train had one and sometimes two cars where all the baggage of the traveling passengers was stored and carried to the next destination. This wagon had four steel-tired, spoked wagon wheels with a rack on the front and the back to keep the suitcases from tumbling off. A handle was attached to the front set of axles and wheels and it could be hooked in an upright position for space-saving reasons. As this wagon was rolled out next to the baggage car, it was an easy transfer of baggage from the wagon to the car, both for arriving and departing passengers.

A washroom for gentlemen and a ladies cloak room and bathroom were on opposite ends of the station waiting area. They had real flush toilets, a novelty for me at that time. We hadn't joined the modern era of indoor plumbing. In fact, we didn't have running water either, a necessity to run flush toilets and water taps. In the first hour of the trip I had observed and experienced a ton of new things, things I had no idea even existed.

Wainwright was a railroad divisional point. That meant that it was a community that serviced the passenger and freight trains that arrived. For the trains it meant a change of crew, a taking on of water for the steam engine, the addition of coal for fuelling the boilers, and a half-hour stop where passengers could get off the train and go to the Beanery, a restaurant at the east end of the station. In addition, Wainwright had a roundhouse, a building built in a configuration that allowed engines to come into a large shed and then be turned around inside and drive out again once maintenance or repairs had been completed. This whole business required a water tower, coal dock, bunkhouse, and a large crew of men to keep it going. Wainwright was an important railway town.

I think our trip to Mission, B.C. was probably made in later March or early April, judging from the early flowers getting ready to bloom when we arrived. If that was the case then the time of the train's arrival was close to the equinox, meaning that as we were catching the morning train it definitely was still dark out. Certainly as we left the waiting room and headed for our coach all that was lit up was the area under the incandescent lights along the platform. There was a hustle and bustle, my hand in Grandma Dixon's and Alison, my younger sister, in my mother's arms. Tickets were grasped by the ladies in their free hands as they looked at the car numbers and then at the tickets until the two numbers matched.

At each entrance point along the platform a conductor or trainman was waiting to assist passengers to board. A bright yellow rubber-topped stool was placed at the entrance where we boarded; with a step on it and then the next stride onto the car stairs, a helping hand from the conductor, and a grasping of the handrail at the side, we were aboard. When we reached the entrance level of the passenger car, we turned in the direction of our car number, pushed open the entrance door, and headed to our seats. We were in a sleeper car, one where you sat up during the day and at night the porter came around and lowered an upper bunk from the ceiling with a very impressive big key and made the lower seats into a bed as well. Curtains came out from their hiding places along the wall and you had your own little enclave to enjoy.

Anyway, we started our journey shortly after boarding. There was a short blast of the whistle from the engineer and then the call up and down the platform from each conductor and trainman, "All aboard!" a short pause and again, "All aboard!" as the stools were lifted and thrown into the area between the passenger cars, ready for the next use at the next stop. Shortly thereafter the train began to move. There was a slight jerking motion as the slack of the car couplers reached their full extent and we were off. Naturally, as we pulled out and began our journey, my nose was pasted to the window to see all the wonders of the countryside that I didn't even know existed.

The journey over the 195-foot-high Battle River Trestle was most impressive and quite scary, as I'd never been higher than the high end of an under-duck on our swing prior to that. This unfolding of the wonder of the prairies continued until we began to enter the forested areas of western Alberta and eventually crossed into the foothills and mountains. We went through Edmonton, where Mom had trained and worked as a nurse a few short years previously. I'm sure that Alison and I slept a fair bit during the time required to get to Jasper. The adrenalin rush of this new adventure couldn't last forever. It was in Jasper that engines were switched to heavier mountain types. These were required to meet the power demands of the steep grades of the road bed that snaked through the mountain passes. I'm sure the adults' sleep may not have come quite so easily as for us two youngsters. What with the passing of other locomotives and trains, the clickity-clack of the wheels over rail joints, the stopping in towns with passengers leaving and boarding, this certainly wouldn't have translated into a restful situation.

I'm sure that the majesty of the Rockies would have grabbed my attention as we passed through them, this being my first experience. The only other thing about the trip I distinctly remember was the trapped deer. As the train rolled along we came to the entrance of a tunnel.

Coming back from the entrance was a page wire fence that probably only extended a couple hundred yards from the tunnel mouth. This, I guess, had been erected to keep wildlife from being injured or killed on the tracks. As we passed I noticed a whitetail buck had entangled his antlers in the page wire and had been there long enough that he was no longer alive. I'm sure the next few miles were filled with question after question that only a four-year-old could think up about what had happened, and what we could do to help the poor deer.

This early experience with rail travel and being next to these massive engines and passenger cars had a profound effect on me. I'm still fascinated with trains. Even now when I happen to be in town with half an hour to kill I will be drawn to the station and the benches on the south side next to the rail. Although much has changed, I'm still engrossed as the large engines pass, pulling literally hundreds of cars behind them. I still marvel at how all this massive equipment, weighing so much, traveling at such speeds, can stay put on two ribbons of steel crossing our great country.

# CHAPTER 28
## The Barnyard

Barnyards are amazing places. They, of course, are made up of the areas around the barn and outbuildings needed to service a farming enterprise. Most often they are surrounded by substantial fencing, not barbed wire with small posts but heavy posts and heavy rails or planking strong enough to withstand working with animals in close quarters. In the case of hogs it was a page wire fence with a strand of barbed wire on the top and bottom. When in use, not a weed or plant of any kind could be found in the well-worn areas, but there were always the weeds that struggled to grow next to a post or in the shadow of the fence rail where the hooves of passing critters didn't trample them. Trampling was one thing; in the hog yard rooting also took place. Hogs naturally use their snouts to "root" through everything. When I was a small boy, I remember our hog lot looking like a well-plowed field. Mixing this good soil with the continual addition of more organic matter that had been run through the hogs made for the blackest and richest soil you could wish for. As you can imagine, when this continued over a period of years you ended up with a fair bit of this material. Interestingly enough, these yards can still be seen even after an old farmstead has been ripped up and put back into crop production. For years the crops grown on this soil are usually superior to those from land right next to it.

Keith Erwin Brower

I certainly remember our corrals and barnyards. In the spring, usually in March, the snow and ice buildup began to melt. It was interesting to watch the cow patties and other darker material sink quickly into the surrounding whiteness as their darker colours attracted the heat of the rising sun. It wasn't long until the snow was gone in these areas and the animal traffic began to mix the winter's accumulation into the soil of the corral. Until the pastures were ready to accommodate the cattle the mixing and trampling continued. Then, as the cattle went to pasture, a few weeds began to try to emerge through the rich soil, except for on the path leading to the barn where the milk cows made their morning and nightly journey. Corners of the corrals would become green with pigweed. This was not all bad, however, as when these plants were young and tender our mother would gather some and prepare them as greens to go along with fresh potatoes, perhaps fresh carrots or peas, and the meat of the day. A little vinegar, Dad's favourite, and they were quite tasty. There was never a shortage at this time of year.

The hog lot was a different story. The reason for the barbed wire at the bottom of the page wire was because of their rooting habits. Without this wire it was common for the group of pigs to root their way right out of the lot. Hogs are smart. Once an escape route had been identified they were very difficult to keep in. They also loved anything green, including the weed named after them. Continually rooting around in their pen soon had everything that appeared green devoured and woe to the weed that tried to push up through the soil there. It was fun to watch them as we small children threw handfuls of anything green: weeds, grass, spent plants, and so on into the pen. There was a free-for-all. There was pushing and shoving, squealing, and general mayhem as the whole herd tried to get their portion first. Sort of reminds me of footage pictures I've seen of today's subway stations in some of the major metropolitan cities of the world. So much can be learned from farm animals when you watch them carefully.

The barnyard was actually a pleasant place to be when things were dry and we hadn't had rain for a while. The slight aromas of horses and cattle were rather subdued at that juncture and as a farm child they were almost comforting. An empty barn without the smell of the occupants just didn't seem right. Not the high-ammonia-type odour but the more subdued aromas typical to any animal shelter. Little breezes that lifted some of the scent from the paddocks left us with knowledge that things were going on here, but were never overwhelming or burdensome to the nostrils.

However, get a three-day rain in the barnyard and all things changed. The yard became a quagmire. Chasing the animals became almost impossible. When they took off at full speed, the chaser became plastered with muck, it was easy to fall, and the dampness brought out the full flavour of all the various commodities that were mixed in the barnyard. I don't recall my mother appreciating seeing us after such an event; however, I don't recall being scolded either. Mom seemed to understand small boys and their adventures. Keep in mind that all the things had to be washed in the old wringer washer and that we didn't have running water in the house at the time.

The composting barnyard materials were also rather soft in areas. I do remember being placed behind the saddle of Mom's palomino, Duchess. She was spirited cow horse and obviously didn't appreciate the extra rider, for I soon found myself in the soft barnyard on all fours, unhurt but a lot lower in the world than I had been a few seconds earlier. I didn't ask to ride there again. The barnyard softness also came in handy when a few years later the Miller boys and us got a few of the steer calves in the chute and put a rope around their middles and tried to ride them. There were not many eight-second rides, as I recall, and it was then and there I confirmed that I didn't have the talent for a budding career in bull riding.

By far the best thing about the barnyard and rich soil areas surrounding it were the mushrooms that emerged after a three-day rain. We boys had a job to do when a period of rain like this hit. These rains were not that common but a couple times a summer they seemed to occur. These were welcome interludes, both for our parents and us children. My brother's and my first task was to sweep the shop. I'm sure Kent did most of the work, being three years my senior, but I was tasked with replacing tools and the like, doing whatever a small guy could do. My father probably had a hard time with the "time crunch" of keeping things running and having to clean up after fixing things. There was a buildup of dirt and materials that even small boys could help with. Once we were finished, the day was ours to do what we liked—until chore time, that is. It was then that we would grab one of the metal calf-feeding pails and, making sure it was clean, go around the outskirts of the barnyard in the search of mushrooms.

There were two varieties that we were looking for, either horse (*Agaricus Arvensis*) or meadow mushrooms (*Agaricus Campostris*). We had been carefully instructed by our father as to what to look for. As the damp soil began to dry, often you would see small cracks appearing as the buttons of these fungi pushed up through the soil. It didn't take them long either. The white caps pushing up through the soil were quite evident. We would land on a patch of these and begin to pull them from the soil. It was very important to check under the cap to look to see if the spore fins were a nice salmon colour. This was critical, as some look-alike mushrooms without this characteristic can make a person very sick and even die. We also had to look to see if small flies had landed on the mushrooms, laid eggs, and the resulting larva had filled the stem and cap with holes and themselves. It was a race, to be sure, flies against boys, and sometimes the flies won. If they had started on the stem, sometimes they hadn't reached the cap yet and we just threw away the stem.

I recall one particular good picking when my brother and I picked over half a pail of beautiful white mushrooms. It was a delight to bring these into our mother and there was no better flavour or treat than to add these fresh mushrooms, fried in butter, to the menu. After a picking like this one, we had mushrooms with every meal for a couple of days.

Chronicles of the Glen

# CHAPTER 29
## A Day of Firsts

An early October snowfall had given partially away to the speck of heat that was still part of the sun's daily trek across the sky. Blotches of bare ground showed through and trickles of melted snow formed puddles and rivulets, only to be refrozen overnight and have the process start all over again. There was no doubt that the whole thing was to end soon; this was an early indicator that winter was on its way. There was a great deal of hustle in our farmyard, for our Dixon grandparents had recently moved off their farm a mile east and had settled a hundred yards through the trees from our farmhouse.

The whole process had not started well. I remember the overheard discussions regarding my grandfather's health. He had been to see doctors; his own son was a doctor. He had a brother who dropped dead bringing up sacks of potatoes in his general store in Hythe. His own father had health problems; these had brought him to the west from New Brunswick in the first place. And so there was the layman's diagnosis that perhaps his heart was bothering him and he needed to slow down. Having trained in Hibbing, Minnesota as a bookkeeper (accountant) and having worked for the Foss Tug and Boat company in Tacoma, Washington, Grandpa did have transferrable skills; skills needed in the fledgling military installation at Wainwright. By today's standards he wasn't an old man, in his mid-sixties I believe, a very knowledgeable and capable person, one inclined to do a job well: a good find for the department at the base where he worked.

I'm sure it was probably a family decision, Grandpa loving the farm but his children realizing that he was no longer able to carry on. It was apparent by now that his son (my Uncle Allen) and daughters, my Aunt Pauline and my mother, were not returning and with his health failing the choice was made to rent out the farm and buy and locate a house on our property. I don't remember the farm sale. I'm sure there was one, where all useable machinery, all remaining livestock, leftover grain and feed, and unneeded household items were sold and turned into cash. There were a lot of these sales every spring in the surrounding country as farms where children had left for better prospects in the surrounding cities did not return to continue in their agriculture heritage. Farms were becoming more mechanized and larger and therefore there wasn't the requirement of farm sites and families on every quarter section. Tractor power was replacing horsepower, and the early onset of electrical power to farms was transforming everything. No doubt some of the funds received from the farm sale were used to purchase and move an empty house from the hamlet of Heath, some fifteen miles to the east of my folks' farm. This house was moved onto my parents' property, a place for my grandparents to live next to our mother, the nurse.

It was that summer that the basement was dug. One piece of machinery that Grandpa couldn't part with was his 8N Ford tractor. He loved this piece of equipment and brought it with him. With it attached to a Fresno (a sort of large scoop shovel with two arms attached to the three-point hitch) and with someone, probably my father, hanging onto the handles that helped control the depth, the tractor was run through the area where the house was to sit. Each time it passed, earth was scooped up and slowly the soil was removed, deep enough to begin the basement. All summer after the hole had been dug, most waking moments were spent getting the concrete basement ready. I do remember being there "helping" (mostly playing in the sand) and always watching the development.

I've seen pictures of others helping, including our two single lady pastors, when the cement was being mixed and the forms being filled. Interesting to see them in pants, or slacks, as they were called. Women weren't often seen without a dress in those early years. A number of other folks pitched in: Grandma's nieces and nephews, Ben and Bernice Miller, railroader Alfred Lilly, and I'm sure many others. Sort of a barn raising in house form.

Once the basement was poured and the forms removed, the next task was to have the house moved from its original location east of Wainwright to its new location in our farmyard. I'm quite sure the moving was done by a couple named Mix; I think Mrs. Mix did the driving. At least I heard a lot about Mix the Movers in those days. They had also moved our small church from Heath into Wainwright a few years previously so I suspect they did the house moving.

I recall seeing the house arrive. In those days you just started out and everyone else stayed out of your way; there were no requirements to get permits, call power companies, and so on. Power wires and telephone wires were lifted and slid along the roof, helped along by someone on the roof (usually my father). The building was slowly moved as the "wire monkey" helped shift the wires along the roof and dropped them back into place as the building passed. These buildings usually had fairly steep roofs, so it was a rather skilled job to stay on top of the roof, move the wires along, and keep from grounding yourself or sliding off. Makes the man-bucket trucks of today look rather appealing.

The reason I remember this so well is that Grandma Dixon's prediction came to pass just prior to the house-moving. For a few months now Grandma had been encouraging us to help our mother all we could. We were to clean up and help with the dishes; even my little two-and-a-half-year-old sister Alison was spurred on to help. Conjecture was that if we didn't work extra hard our mother could "end up in the hospital." This seemed to us worse than a sentence to death, for whatever reason. We seemed to think it was the equivalent of being sent to prison or some similar fate. And it happened; my mother ended up in the hospital. It seemed like a real tragedy until it was announced that my second-youngest sister, Bethany, had been born. As a four-year-old I wasn't aware that Mom had put on weight, but when she hugged me she was a little further away. The change had been so subtle I hadn't even noticed. I was very pleased that she hadn't been shipped off to some dreadful place and was reassured that in a few days she would return. So it hadn't been slack work habits that had sent her to the hospital after all.

All during that summer construction continued. It had to be fit in between all the farm work, including springtime and harvest, livestock feeding, and milking. By today's standards it wasn't a huge job but when most of the work was done by hand, including concrete mixing, it was a substantial project. Of course, as with many projects that are done in the cold climate areas of the planet, the whole thing was spurred on realizing that most of this work had to be done before the fall winds brought on the onset of winter. There was much to do before then.

On an October day in 1954, a few days after October 14th, the house arrived from Heath to be set on the prepared foundation and the time arrived for my new sister to come home. I distinctly recall looking out the window in the fading daylight as the house was moved onto its new resting place, ready to start a new life with the new family that would soon occupy her.

I left my post at the window and stole away to the other fascination of the day, found on the front porch by what we called the front door, and carefully peered over the side of the baby buggy. There slept a tiny little girl with a hint of dark hair and tiny hands and fingers. She too had arrived to start a new life with her new family, in her new home. Such a busy day of firsts for all of us.

# CHAPTER 30
## Mom and Dad

My father and mother were poster people when it came to the theory that opposites attract. My mother was reserved, didn't like the spotlight, and was a quiet and a behind-the-scenes type of person. This didn't mean that she was "soft" or wishy-washy with no backbone and couldn't hold her own. Oh yes, she could; she could get more done in a day than many of her compatriots and have time left over to run her unofficial non-paying counselling service. Her caring, timely advice and soft heart were a perfect combination for some of the folks with life difficulties to prevail upon her listening ear. It was also the perfect personality for her to become a registered nurse after her high school career. Many nights Mom would be on the phone with one or more of these folks, listening to their problems, giving advice, but mostly just being there. Her "clients" ranged from neighbours to cousins and other relatives, all needing something. Often Mom was the one called to be a special nurse for dying neighbours, cousins, uncles, and aunts. She was there on many occasions to ease the passing of these mortals into the next world. I don't think the group often thought of the commodity Mom needed most, with five kids of her own, eventually: time. At any rate, Mom was in high demand.

I remember one of these occasions when I was involved as my grandmother's sister, Aunt Edith Woodward, was gravely ill (I assume with cancer). My job was to count for Aunt Edith. She obviously enjoyed small children. I recall being brought into her bedroom and immediately noticed a white head on a pillow and the completely covered frail body of older lady lying there. She asked if I could count for her. Naturally I was game and began my counting. As I recall, the first ten numbers went very well; the next, well, they were a complete hodgepodge but Aunt Edith patted my hand and gave me encouragement after I had recited every number I could think of. It brought a smile to a face that I'm sure really didn't have a whole lot to smile about most days. At any rate, in a way, I had played a small part in Mom's role as a giver of solace to so many.

My father, on the other hand, was an extrovert. Dad was nearly always singing or whistling as he worked. It was a good way to find him as he wandered around the farmyard. Just stop and listen for awhile and you could home in on his location with great accuracy. Sort of like a very early GPS system. If you knew that Dad was in the yard and not whistling, it was clear that something weighed heavily on his mind, or it could mean that there was upcoming discipline for some overlooked chores or some other misdemeanour. Sometimes it paid to give a wide berth, in this case. Dad wasn't unfair, but he did require unwavering obedience, and we learned very early on to obey. If a job was worth doing, it was worth doing well.

I think that you could say that Dad generally liked people. I'm not sure he ever met anyone he didn't like or couldn't at least get along with. Dad also had time to enjoy the wonder of the world he lived in. He fed the birds, learned how to accurately make their calls, observed the habits of all the wild animals found on the farm, and generally took pleasure in all the natural surroundings that were readily at hand in this environment. From an early age it was fascinating for us as children to observe all the birds and their characteristics. In spring we listened for the drumming of the rough grouse, the call of the killdeer, the meadowlark, ducks and geese. Dad's old bird book was well-thumbed by him and us children as we tried to identify the latest spotting.

It was almost humorous when Mom and Dad's different life views crossed. Dad would call, "Jeanie, come over here and look at this bird at the bird feeder."

Mom would say, "I know, Barry, I've seen it before."

"But this one looks like it slept in a round hollow log; its tail is curved."

Mom would reluctantly leave the kitchen counter, have a quick glance, say, "I see," and quickly return to her work.

Another example of how they viewed the world differently was the conversation that would follow a children's concert. It went like this. Dad would say, "Wasn't that a great night, with all the kids there in their finest, singing and having a good time?"

Mom would say, "That was such a poor concert; there were so many kids that didn't know their lines, they were noisy, they didn't speak out very well, and the bigger kids were in the wrong spot, you couldn't see some of the little ones." It was sort of a glass half-full or glass half-empty scenario.

This certainly didn't hamper the team they developed into over the years. They were very supportive of each other's strengths and helped each other cope in the area that they were not so strong. Dad's love of flowers and animals did begin to rub off on Mom. Her sense of order and neatness combined with Dad's love of flowers turned many a drab corner around the farmyard into a beautiful flower garden, with colourful shrubbery and perennial beauty. The neighbour ladies of the district, who gathered once a month at the Sydenham-Gerald FWA (Farm Women of Alberta) meetings, traded plants and seeds. Irises, lilies, bleeding hearts, Virginia creeper, the list went on and on. They were all willing to share with each other, something that rural neighbours are known for. And so many of these plants became an important part of our summer floral displays.

Mom's exceptional neatness and organization helped Dad throughout the years as well. She was the one, with a child or two in tow, that went down to the barn to wash the cream separator, gather the eggs, milk the cows on occasion, and execute any other duties that required her hand.

Dad was not messy. He loved to have a perfectly straight well-built fence, a well-designed barn, and well-maintained equipment, for example. He would never leave old posts attached to barbed wire in any fence. The old posts were removed and the new ones spaced exactly twelve feet apart with each one standing straight and true. He had an eye for architectural beauty as well: a farmhouse roof that looked just right with proper eves, a barn that was in proportion roof to walls with just the right angle on the roof pitch. It was just that sometimes he had so much on his plate that he had to leave things where they were once the item he had been fixing was functional. It was amazing how this young couple kept it all together having so much to do.

Sunday afternoons were quite often set aside to go for family walks. More so in the warmer months than the winter. We did acquire an appreciation for the beauty of winter as well, however. We just normally didn't go for long Sunday afternoon walks in those colder times. We gleaned from both our parents the ability to look at the things around us and enjoy them. There was the beauty of drifted snow, for example. As snowflakes and crystals were moved across the land by the action of the wind, wonderful patterns were sculpted into the surface. With the bright sunshine on a calm winter's day combined with the slant of the sun's rays, the etched snow showed patterns highlighted by blue in the silhouetted areas, making outstanding works of art. Other winter wonders became apparent as snow drifted down on the trees and grasses around our area. Hoarfrost made the trees seem almost magical as crystals formed from the moisture left by low-lying clouds that had hung around for a few days. Although all these wonders were appreciated, it was when the days grew longer and the snow began to melt that we were the most excited, for we knew the Sunday afternoon walks would commence for the summer. It was here that Dad's nature interest peaked. Quite often Mom accompanied us but not necessarily for the long haul, as she would return to the house to listen for the baby rustling after her afternoon nap.

Over the summer we would scour the many trails and bluffs behind the farmyard. It was here that we became acquainted with the flora and fauna of the area. We knew about the mint that grew in the partial shadow of the treed areas. Dad showed us how to pick the leaves, look for the heart of the leaf in the stem, and pull that heart out, which gave us a woody sort of strand to chew on for a mint flavour. There were the wild tiger lilies that could be found in the same environment and the wood violets. Early on in the season there were buffalo beans with striking yellow pea-like flowers; even the dandelions as they began to bloom would become a pretty yellow ground cover. In the marshy areas there were shooting stars, beautiful lady slippers, and bog violets and buttercups. Some of these we were not allowed to pick because of how rare they were; picking would make it so that these individual plants would not survive. As the summer wore on the early bloomers were replaced by such things as alfalfa, clovers, and Canada thistle. Finally, into fall the Black-eyed Susans, the goldenrods, and other such beauties lent their colour and fragrance to the whole locale.

It was, however, the early spring and the arrival of the prairie crocus (Pulsatilla patens and Anemone patens) that was the most captivating for all of us. My brother and I, along with our parents and my sister Alison, would be out on the first warm Sunday in spring to see if we could find this early herald of spring. Find a nice south-sloping patch of prairie wool, one where the snow had disappeared, and usually if any amount of warmth and sunlight had been directed onto it, the buds of some and the full growth of others of the first flower of spring would be present. I remember a whole hillside being blue some years with these crocuses as they bloomed.

There we would all be, crawling around between these beautiful little blue flowers. They were such a joy to behold, such a proclamation that winter had left us and spring was on the way. To me, since I was on my knees anyway, it was a natural time and place to raise my head and say thank you to the creator that the winter was past and to declare that the beauty of my surroundings were so much appreciated.

Printed in Canada